Name: ....

MW00610355

"JAIN 108 MATHEMAGICS CURRICULUM
For The GLOBAL SCHOOL, Book 4"

subtitled:

Workbook Level 2

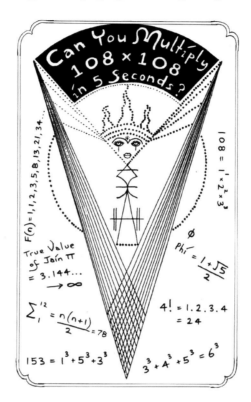

RAPID MENTAL CALCULATION
for TEENS and ADULTS

Designed to release the Inner Genius

(Script Material for the Film and OnLine Tutorials).

By JAIN 108

(writer, artist, **Mathematical Futurist** and EcoPreneur)

2011

Produced by JAIN F.R.E.E.D.O.M.S
(For Research Expressing Essential Data Of Magic Squares)
www.jainmathemagics.com

ISBN: 978-0-9757484-7-3

Level 2
MATHEMAGICS for TEENAGERS
2011

CERTIFIED CERTIFICATE COURSE
10 Afternoons from 4pm to 6pm

CONDUCTED by JAIN

CURRICULUM NOTES
Presented as Worksheets
As a companion Manual for the upcoming
5 DVD Set from DAY 1 to DAY 10

This booklet is a comprehensive LIST of **WORKSHEETS**
to be GIVEN to the STUDENTS
nb: This booklet is also a GUIDELINE or MANUAL for
TEACHER TRAINING SEMINARS

This COURSE REPEATS EACH SEASON or TERM
in any village, school, institution or country of the world

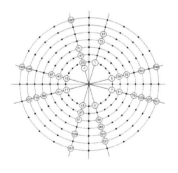

BEGINNING Of
RAPID MENTAL CALCULATION
LEVEL TWO
(JAIN MATHEMAGICS ONLINE CURRICULUM
For The GLOBAL SCHOOL)
For TEENAGERS From AGES 13 To 19

JAIN MATHEMAGICS CURRICULUM
for the GLOBAL SCHOOL

formerly known as: The VEDIC MATHEMATICS CURRICULUM for the GLOBAL SCHOOL

$$13 \times 14 = 13 + 4 / 3 \times 4 = 17 / 12 = 18 / 2 = 182$$

$$3^3 + 4^3 + 5^3 = 6^3$$

$$98 \times 97 = 98 - 3 / 2 \times 3 = 95 / 06 = 9,506$$

Part 2 MULTIPLICATION   JAIN 2005
Developing the Inner Mental Screen

Jain: International Lecturer on Vedic Mathematics and Sacred Geometry),
Mathematical Futurist and EcoPreneur (not entrepreneur).
In the background on the cloth banner is a
4x4 magic square rotated upon itself 4 times

CONTENTS

INTRODUCTION

MAKING A NOTE OF THE SUTRA OR SUB-SUTRA EMPLOYED

After the completion of each day lesson, make a summary of what were the major sutras or formulae taught and make a note of this by referring to the chart that lists both
1 – **16** Vedic Mathematics **Sutras**
2 – **14** Vedic Mathematics **Sub-Sutras**.

According to the previous DAY 1 Lesson, the main sutra covered was called DIGIT SUMS or DIGITAL COMPRESSION but was known 5 decades ago as
THE SUM OF THE PRODUCTS, and its Sanskrit name is:
SAMUCCAYA-GUNITA.
This sub-sutra was used in the Multiplication of 11, as in 25x11 = 2/**2+5**/5 = 275 where digits were added to achieve an instant answer.
The teacher must now guide the students to mark this special sub-sutra by referring to the page that says: "Sub-Sutras" and ticking the number "10" that refers to "The Sum of the Products".

You can progressively mark your Sutras with a tick or an arrow or any other creative squiggle that shows you that you have read and understood the formula.

LEGEND:

aka = also known as
LHS = Left Hand Side
RHS = Right Hand Side
ibid = same as previous listing
∴ = therefore
ie: = that is
eg: = for example

CHART OF THE 16 MAIN AND 14 SUB-SUTRAS

16 SUTRAS	15 SUB-SUTRAS
1 - By One More Than the One Before	1 - Proportionately
2 - All from 9 and the Last from 10	**2 - The Remainder Remains Constant**
3 - Vertically and Cross-wise	**3 - The First by the First and the Last by the Last**
4 - Transpose and Apply	4 - For 7 the Multiplicand is 143
5 - If the Samuccaya is the Same it is Zero	5 - By Osculation
6 - If One is in Ratio the Other is Zero	**6 - Lessen by the Deficiency**
7 - By Addition and by Subtraction	**7 - Whatever the Deficiency lessen by that amount and set up the Square of the Deficiency**
8 - By the Completion or Non-Completion	8 - Last Totalling 10
9 - Differential Calculus	9 - Only the Last Terms
10 - By the Deficiency	**10 - The Sum of the Products** aka **Digit Sums** or **Digital Compression**
11 - Specific and General	11 - By Alternative Elimination and Retention
12 - The Remainders by the Last Digit	**12 - By Mere Observation**
13 - The Ultimate and Twice the Penultimate	13 - The Product of the Sum is the Sum of the Products
14 - By One Less than the One Before	**14 - On the Flag**
15 - The Product of the Sum	**15 – Trachtenberg** 2Digit x 1 Digit
16 - All the Multipliers	

(nb: I have added the 15th Sutra: "Trachtenberg" to honour this discovery).

Notice how the relevant **10th Sub-Sutra: "DIGIT SUMS"** has been highlighted.

Of these 30 formulae, 13 have been marked in bold and they are the actual ones that will be dealt with in this 10 week course according to the teachers time management in the School Term.

Level 1 (One) Course is called **Beginners**,
then subsequent courses like
Level 2 (Two) is **Intermediary** and
Level 3 (Three) is **Advanced**
will deal with most of the others sutras.

Do not worry about these complicated names of the Vedic Sutras, they will all appear to make sense by the end of the course.

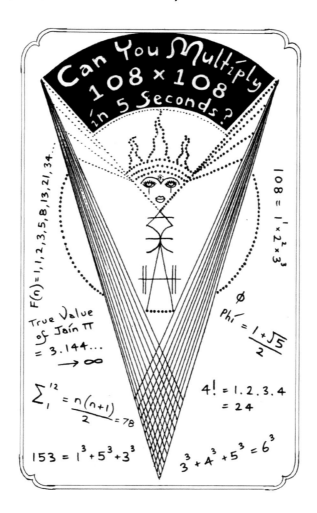

Beginning of course

Registration: Date ....

Venue: At ....

Name Introductions by throwing ball around: stating :
- My name is .... and my favourite fruit is ....

   This also teaches Memory Power by using an association with an Image.
- **Introduction** to Vedic Mathematics and Pattern Recognition

- **Revision** of the Days work.

- **Kumon** = Japanese concept of revising material within 24 hours to lock the information in your brain for life.

- **nb**: These lectures are leading to all day seminars on Sacred Geometry eg: ATOMIC STRUCTURE Of PLATINUM CRYSTAL BASED On The DIGITAL COMPRESSION Of The MULTIPLICATION TABLE. Thus we introduce the Times Tables in this 10 week course, but during the school holidays, we go deeper into full day studies.

Other exciting topics include:

† The Prime Number Cross
† The Binary Code
† Magic Squares,
† 5 Platonic Solids,
† Phi Ratio...

DAY 11

PASCAL'S TRIANGLE

- **Theory**. Get students to create this Triangle and discover the following 6 hidden Sequences:
- **Binary Sequence**   1-2-4-8-16-32
- **Powers of 11**   1-11-121-1331-14641
- **Sequence of Triangular Numbers**   1-3-6-10-15-21
- **Sequence of Tetrahedral Numbers**   1-4-10-20-35
- **Fibonacci Sequence**   1-1-2-3-5-8-13-21-34
- **Fractal Pattern of the Odd Numbers**

Blaise Pascal

```
                1
              1   1
            1   2   1
          1   3   3   1
        1   4   6   4   1
      1   5   10   10   5   1
    1   6   15   20   15   6   1
```

EXERCISE 1

On this page, draw your own Pascal's Triangle.

Add up all the digits of each horizontal row (R).
What are their sums?

R1 = 1

R2 = 1+1 =

R3 = 1+2+1 =

R4 =

R5 =

R6 =

R7 =

Is there anything special about this Sequence of Numbers?

EXERCISE 2

What is special about each row when viewed as a number:
Clue: Think of the number "11".

Row 1 = 1 =

Row 2 = 11 =

Row 3 = 121 =

Row 4 = 1331 =

Row 5 = 14641 =

Row 6 =

How can R6 be simplified so that it is expressed as a Power of 11.
Using the "Vedic" tradition for expressing numbers:
Insert "Forward Slashes" between the numbers, and allow only 1 digit space between the partitions, therefore moving any "carry-overs" to the LHS.
eg: $1/6/_15/_20/_15/6/1$

Draw a rectangular Frame around the sequence of numbers running downwards in this fashion: 1-3-6-10-15- etc
This Sequence is known as the **Triangular Numbers** (TN) which is really the sum of counting numbers in their natural consecutive order eg: ("Consecutive" means 1-2-3-4-5-6 etc).

TN1 = 1 =

TN2 = 1+2 =

TN3 = 1+2+3 =

TN4 = 1+2+3+4 =

TN5 = 1+2+3+4+5 =

TN6 = 1+2+3+4+5+6 =

SIGMA FORMULA FOR THE TRIANGULAR NUMBERS

If I wanted to predict the next number in this sequence TN7, there is a special algebraic formula which says we take this number of 7 (call it n=7) and multiply it by the next counting number which is 8 (call it n+1=8) then half the product or answer. So instead of counting (1+2+3+4+5+6+7) we use an intelligent formula based on Sums, called Sigma, a Greek symbol written as (Σ). We say that

"Sigma 1 to 7" = (7x8) ÷ by 2 = 28

$$\sum\nolimits_{1}^{7} = \frac{7 \times 8}{2} = 28$$

As a general formula for the Triangular Numbers, where n=number, learn this:

Sigma Σ 1 to n = n(n+1) ÷2

$$\sum\nolimits_{1}^{n} = \frac{n \times (n+1)}{2}$$

Draw a rectangular Frame around the sequence of numbers running downwards in this fashion: 1-4-10-20-35- etc

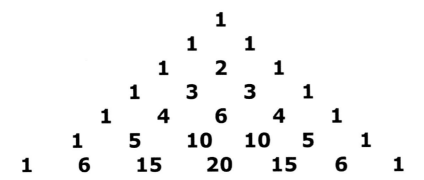

```
                    1
                 1     1
              1     2     1
           1     3     3     1
        1     4     6     4     1
     1     5    10    10     5     1
  1     6    15    20    15     6     1
```

This Sequence is known as the **Tetrahedral Numbers** (TetN) which is really the 3-dimensional version of the Triangular Numbers where we are adding consecutive triangular based Pyramids of numbers.

ANECDOTE:

HOW GAUSS SOLVED THE FORMULA FOR THE TRIANGULAR NUMBERS

There is a story told of a great German Mathematician Karl Gauss, who solved a problem at the age of 10, in a very smart way, not with the traditional formula, nor with the Vedic Math Sutra. (See Fig 29 below). When his teacher asked him to add 1 + 2 + 3 + 4........ + 18 + 19 + 20 Gauss quickly gave the answer. He had 3 lines of setting out:

1)   1 + 2 + 3 + 4 + 5 + 6 + 7 + 8 + 9 + 10 + 11 + 12 + 13 + 14 + 15 + 16 + 17 + 18 + 19 + 20
2)   20+19+18+17+16+15+14 +13+12+ 11 + 9 + 4 + 8 + 7 + 6 + 5 + 4 + 3 + 2 + 1
3)   21+21 +21+21+21+21+21+21+ 21 + 21 + 21 + 21 + 21 + 21 + 21 + 21 + 21 + 21 + 21

Fig 29
Gauss' Method of Adding Numbers

Gauss wrote the second line the same as the first line but in reverse order. This makes apparent that the third line is 20 pairs of 21 or 20 x 21. Thus, the desired sum is half of this or (20 x 21) ÷ 2 = 210. (or algebraically = [n x (n + 1)] ÷ 2 (Remember, this is the formula for the Triangular Numbers).

This account is merely to demonstrate that when we don't know a specific formula, there are many ways of approaching the problem, not always in the way we expect to solve an operational problem, but most successfully when we dissect the data with the keen X-Ray eyes of a Pattern Hunter.

EXERCISE 29

To continue your practice with the multiplication of your Teen Numbers, what are the Sums or Sigma of the following Triangular Numbers:
a) TN 10   **b)** TN 13   **c)** TN 14   **d)** TN 15   **e)** TN 16   **f)** TN 17   **g)** TN 18   **h)** TN 19   (see answers on p90).

The above diagram is taken from my main book on Vedic Mathematics called: JAIN MATHEMAGICS CURRICULUM FOR THE GLOBAL SCHOOL: Book 2: Multiplication. Sub-titled: Developing the Inner Mental Screen. By Jain 2005

MANY SEQUENCES IN ONE MATRIX

("**Matrix**" is a Latin, ancient Roman word meaning "Womb", in the sense that this Pascal's Triangle gives birth to many other patterns). Here is another diagram below showing most of the Sequences discussed. Pascal's Triangle is also known as "The Chinese Triangle" and in India, around 795AD as Halayudha's Triangle).

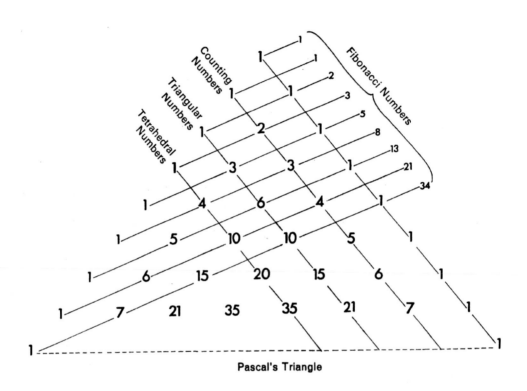

Pascal's Triangle

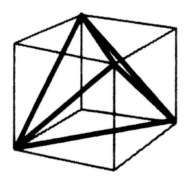

(View of the Tetrahedron inside the Cubic Form)

Here are some views of Tetrahedral Sphere Packing that generate the Tetrahedron Numbers:

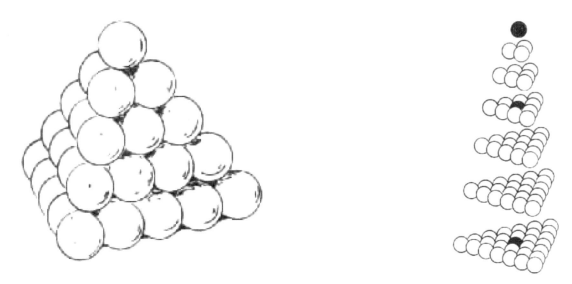

Above:(typical cannonball packing is tetrahedral)

Below: Here is visual image of these Tetrahedral Numbers relating to Nuclear Physics:

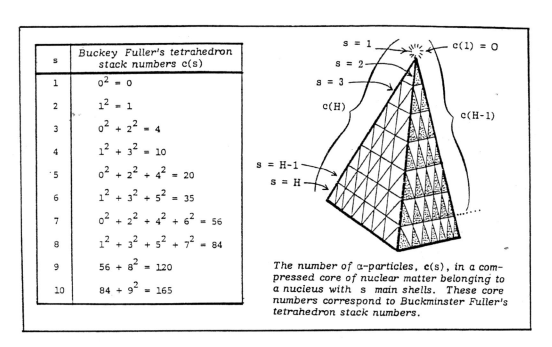

s	Buckey Fuller's tetrahedron stack numbers c(s)
1	$0^2 = 0$
2	$1^2 = 1$
3	$0^2 + 2^2 = 4$
4	$1^2 + 3^2 = 10$
5	$0^2 + 2^2 + 4^2 = 20$
6	$1^2 + 3^2 + 5^2 = 35$
7	$0^2 + 2^2 + 4^2 + 6^2 = 56$
8	$1^2 + 3^2 + 5^2 + 7^2 = 84$
9	$56 + 8^2 = 120$
10	$84 + 9^2 = 165$

The number of α-particles, c(s), in a compressed core of nuclear matter belonging to a nucleus with s main shells. These core numbers correspond to Buckminster Fuller's tetrahedron stack numbers.

(image taken from Chris Illert's book on Alchemy)

And last but not least is the humble yet profound **Fibonacci Sequence** 1-1-2-3-5-8-13-21-34 to key to all biological systems like pine cones and sunflower florets and the mathematics of human limbs like "where the elbow bends" and planetary distances from the sun.

Add all the diagonal rows highlighted to check that they have summations identical to the Fibonacci Sequence.

Sunflower with 21:34 counter rotating spirals. Computer Graphic.
(in all my research, this image above, is perhaps the most psycho-active I know of, and deserves to be used by the student in some creative means).

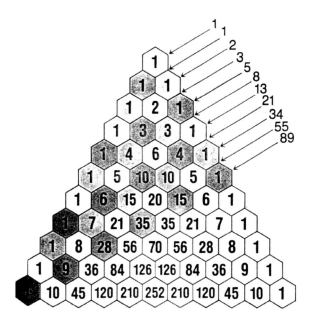

Fibonacci Numbers in Pascal's Triangle

To see this image in another way, look what happens when we rearrange Pascal's Triangle so that all the rows are now Left-Aligned!

The Diagonals sum to Fibonacci numbers?

This time I am going to highlight two diagonals in bold and another diagonal underlined (1 – 4 – 3) so that you can begin to see the first pattern:

```
1
1    1
1    2    1
1    3    3    1
1    4    6    4    1
1    5    10   10   5    1
1    6    15   20   15   6    1
```

Fig

Highlighting the Diagonals to reveal the Fibonacci Sums

Can you see in bold that 1 + 3 + 1 = **5**

Can you see underlined that 1 + 4 + 3 = **8**

Can you see that in bold 1 + 5 + 6 + 1 = **13**

The numbers of the diagonals continue like this forever! Always adding to a Fibonacci Number, in order.

• Fractal Pattern of the Odd Numbers

The **Sierpinski Curve** is generated from the subdivision of the Triangle at midpoints within the Triangle:

Look what happens now when we highlight all the Odd Numbers in Pascal's Triangle!

This modern design has great use in the Mathematical field of Probability.

DAY 12

PERCENTAGES

- Examples like "What is 30% of 370?"
- Wholesale Book Price
- How Algebra Can Be Used To Convert Fractions to Percentages.
  ~ Introduction of "x" The Unknown.
  ~ Convert the fraction 7/40 as a Percentage

The word "PerCENT" is a relationship to 100, the Latin or ancient Roman word "Cent" meaning One Hundred.
Thus when a percentage arises, write it as a fraction referenced to 100.

EXAMPLE 1

What does 20% mean ?

eg: 20% means 20 parts per 100 or 20 ÷ by 100 or "20 over 100" or simply 20/100 or

$$\frac{20}{100} \text{ can be simplified to } \frac{2}{10} \text{ (and decimalized as .2)}$$

The following problems on percentages are designed to revise the previous sutras learnt in the first workbook Level 1

EXAMPLE 2

What is 30% of 370 ?

First step is to find out what 10% is or 1%.
Visualize 10% in its fractional form of 10/100 or simplified as 1/10.
We therefore multiply 370 by 1/10.

$$= \frac{370}{10} \qquad or \quad \frac{30 \times 370}{100}$$

Crossing out the zeroes top and bottom gives 37.
But this is only 10%, we require 30%, so we need to multiply 37x3.
Do you remember the sub-sutra for multiplying a 2 Digit Number by a 1 Digit Number?

It's the Trachtenberg Method.
37x3 = 3x3/3x7 = 9/21 but this is not the answer as we are only allowed 1 digit on the right hand side of the forward slash, so we slide the "2" of "21" to the left and thus the answer is 9+2/1 =11/1 = 111.
Therefore 30% of 370 = 111

EXAMPLE 3

What is 25% of 96 ?

We know that 25% is really 25/100
but when simplified it is really ¼.

$$= \frac{1 \times 96}{4} \qquad = 24$$

EXERCISE 1

Determine the following Percentages:

a) 7% of 3,300   =

b) 11% of 12,300   =

c) 9% of 4,800   =

d) 35% of 3,500   =

e) 23% of 2,700   =

f) 66% of 4,600   =

g) 98% of 9,700   =

h) 13% of 1,400   =

i) 54% of 5,400   =

j) .618% of 34   =

k) Wholesale Book Price (see next page)

a) 7% of 3,300   = 7/100 x 3300 = 33x7 (Use Trachtenberg) = $21/_21$ = 23/1 = 231

b) 11% of 12,300= 11/100 x 12300 = 123x11 = use format 01230 for multiplying by 11 using sutra: "Add The Neighbour" = 1,353

c) 9% of 4,800   = 9/100 x 4800 = 48x9 (Look at the Back of your Hands, make a "V" after the 4th Finger and bend the 8th Finger) = 432

d) 35% of 3,500   = 35/100 x 3500 = 35x35 (Use Sutra: "By One More" For The Squaring Of Numbers Ending In 5) = 3x4/5x5 = 12/25 = 1,225

e) 23% of 2,700   = 23/100 x 2700 = 23x27. (These numbers are cognate since the Tens Column digits are similar and the Units Column digits complement to 10, so we apply the sutra: "By One More") = 2x3/3x7 = 6/21 = 621

f) 66% of 4,600   = 66/100 x 4600 = 66x46 (Use sub-Sutra: "The First By The First & The Last By The Last" since the Tens Column digits complement to 10, and the Units Column digits are similar) = (6x4)+6/6x6 = 30/36 = 3,036

g) 98% of 9,700   = 98/100 x 9700 = 98x97 (Use Sutra: "By The Deficiency" using Base 100) = 98-3/2x3 = 95/06 = 9,506

h) 13% of 1,400   = 13/100 x 1400 = 13x14 (Use Sutra: "By The Excess" using Base 10) = 13+4/3x4 = $17/_12$ = 18/2 = 182

i) 54% of 5,400   = 54/100 x 5400 = 54x54 (Use Formula: "Squaring of Numbers In The Fifties" = $25+4/4^2$ = 29/16 = 2,916

j) .618% of 34   = 618 x 34 then move the decimal point 3 spaces to the left. (Use Sutra: "Vertically & Crosswise" for 3 Digits x 3 Digits), thus multiply 618 x 034 = 21,012 then move the decimal point 3 spaces to the left = 21.012. (34 is a Fibonacci Number, and the preceding number is 21).

k) REAL LIFE CALCULATION IN A BOOKSTORE:

I went to a bookshop to sell this book, hot off the press, that retails @ $44.00 The shopkeeper liked it and asked for a box full of 25 books. The standard wholesale price (WP) is a 40% deduction from the Recommended Retail Price (RRP).
1 – What was the WP per book?
2 – How much was the total of the cheque made out to me, for the total value of the box of 25 books?

1 – We need to know what 40% is of $44.00 for the wholesale price of each individual book. = 40/100 x 4,400 cents or 40 x 44 = (44x4) x 10.

To calculate the first part (44x4) use Trachtenberg Method = 16/16 = 17/6 = 176 and times this by 10 gives 1,760 cents or $17.60

This amount of $17.60 is the Wholesale deduction of 40%, so we need to subtract this amount from the $44.00 to determine the actual Wholesale Price.

Thus $44.00 - $17.60 = $26.40

Therefore, the WP per book is $26.40

2 – The value of the cheque is therefore this wholesale price of $26.40 multiplied by 25. Since is difficult to multiply $26.40 x 25 (using the sutra "Vertically and Crosswise" for 264 x 025), a shorter or quicker way is to think of 25 as 100÷4. This sutra at work here is called "Transpose and Apply"

= ($26.40 x 100) ÷ 4

= $2,640 ÷ 4

= $660

nb: The beauty of this question lies in the fact that the Vedic Mathematician or Rapid Mental Calculator is able to do all of the above calculations swiftly and effortlessly in their head, competing as it were with the shopkeeper who is using their electronic calculator. It's pure delight when we are able to mentally work out the final answer quicker than the shopkeeper, surprising him or her with our mental acuity.

It's not that we are against the use of the calculator, we praise the gifts of our ever-growing technologies, its more the over-dependence of the electronic gadgets that train our brain to become lazy through lack of use. Our brain is a Mental Muscle, it simply needs to be exercised on a regular basis, and as the decades roll by, it will become increasing obvious that if we don't use our mental faculties, as in the above real life calculation, the global brain will atrophy, suggesting that we think we are so intelligent in this new age of gizmos and devices yet ironically we are becoming dumber and dumber! Like der!

HOW ALGEBRA CAN BE USED
TO CONVERT FRACTIONS TO PERCENTAGES.
INTRODUCTION OF "x" THE UNKNOWN

EXAMPLE 4

Convert the fraction 7/40 as a Percentage

There are two methods.
Method A is the simpler way
And Method B is the longer way but is a good example of Algebra.

Method A
7/40 expressed as a Percentage:
Just multiply the numerator by 100
= (7 ÷ 40) x 100
= 400 ÷ 7
= 17.5%

Method B

Now the beauty of Algebra is that we take something that is UNKNOWN, and call it "x", and with a few steps, it becomes KNOWN. We can summarize and state that the Indian or Vedic Mathematics, from 2,000 years ago was a brilliant and amazing system of rapid mental calculation, but when the Arabs raided India, and introduced Indian Maths into the Middle East and Europe, they also are credited with advancing the world of mathematics by introducing the concept of "x" the Unknown and calling it Algebra.

Here is a good example of how we do this:
We ask the question: " The fraction (7 ÷ 40) is the same as "something" divided by 100. We will call this "something" or this "unknown" a symbol called "x"

We therefore call write down the following equality

$$\frac{7}{40} = \frac{"x"}{100}$$

We now "Cross-Multiply" which means we multiply the numbers that are diagonally opposite one another:

7 x 100 = "x" x 40

Simplifying
700 = 40"x"

We can rearrange the equation, as we want "x" on the LHS
40"x" = 700

"x" = 700 ÷ 40
    = 17.5%

Thus we started with an Unknown, and resulted in an "Known"
This is the genius of Algebra!

Using ALGEBRA
to find a number when a percentage of it is known.

QUESTION:

**Lily pays $225 tax per month
which is 15% of her income.
What is Lily's income per month?**

ANSWER 2

15% of "x" = 225

15% is really $\dfrac{15}{100}$

15% of "x" is $\dfrac{15"x"}{100}$

∴ to express this algebraically, where "x" is the Unknown we set up this equality:

$\dfrac{15"x"}{100} = \dfrac{225}{1}$ then cross-multiply
                (nb: 225 is the same as the fraction 225÷1)

$15"x" = 22500$

$"x" = \dfrac{22500}{15}$ nb: What is the Square Root of 225, ie: what number multiplied by itself gives 225?

$"x" = \dfrac{15 \times 15 \times 100}{15}$

$"x" = 1500$

∴ Lily earns $1500 per month.

An ALGEBRA PROBLEM Relating to RATIO

EXERCISE 3

QUESTION:

On a house plan, the length of a room is shown to be 30mm, and the actual room being built, is to be 3.24m long.
What is the ratio of the plan length to the actual building?

ANSWER 3

First thing to do is to express all measurements in the same format. Express all measurements in millimetres (mm), the ratio of 30mm and 3,240mm.

We have the algebraic equality:

30"x" = 3240   divide both sides by 30 to give one "x"

"x" = 108

∴ the ratio is **1:108**

DAY 13

FRACTIONS

- Adding and Subtracting Fractions
  eg: 1/3 + 2/7   and 6 and 4/5 – 2 and ¾
  Understanding Vulgar Fractions eg: 1½ = 3/2

- Multiplying and Dividing Fractions
  eg: ½ x ¼   and   5/6 ÷ 2/5

- Converting Fractions into their Decimal Equivalents

- Non-Recurring or Non-Circulating Decimal Fractions
  When 2 or 5 or 10 is a factor of the denominator
  eg: 1/8   and 1/25 and 1/40

- Fully Recurring or Fully Circulating Decimals
  When 3 or 7 or 11 are contained only in the denominator
  eg: 1/3 = .333333 repeater   and 1/9 and 1/11 and 1/7
  Creating "The Circle or Wheel Into 6 Divisions" to solve 2/7 etc
  eg: 1/7 = .142857 Repeater + Number Theory

- Drawing the UniCursal Hexagram

CONVERTING FRACTIONS
INTO THEIR DECIMAL EQUIVALENTS
USING MENTAL CALCULATION

It is a good practice to know "off by heart" all the simple fractions in their decimal forms. In modern days, it is necessary to know your Decimal Coinage for international trade, as well as Decimal Weights and Decimal Measurement.

Practising Mental Calculation has extreme importance in developing the students Mathematical Confidence and Memory Power.

The Indian Vedic culture, several thousands of years ago, invented this Base 10 application to number crunching, by developing the concept of The Zero, The Decimal Point and the Place Value System, without which we, as a planet, would still be in the Dark Ages. This decimal system advanced planetary consciousness to the next level or octave of awareness.

NON-RECURRING
or NON-CIRCULATING DECIMAL FRACTIONS:

There is no recursion or repetition in the decimal of a fraction to be converted when there is **2** or **5** or **10** as a factor of the denominator (or bottom part of the fraction).

EXAMPLE 1

What is ½ ?

Simply divide the Denominator by the Numerator,
ie: The Top By The Bottom
[nb: here 2) 1.000 is the same as: 1.000 ÷ 2 or simple division]

½ = 2) 1.000   = .5

which is arrived at by understanding that 10 is divided by 2.
Add as many zeroes as required until there is no remainder in the divisions.
For those students not knowing their simple division, you can start by saying 2 into 1 does not go, so put down a zero. Then place the Decimal Point exactly above the original Decimal Point. Then say 2 into 10 goes 5. Giving an answer of 0.5

EXAMPLE 2

What is ¼ ?

¼ = 4) 1.000
or simply half of the previous ".5" . Think of the .5 as .50, thus half of 50 is 25, thus = .25
or ¼ = ½ x ½ or = .5 x .5

nb: With simple division, we say 4 into 1 does not go; then 4 into 10 goes 2 with a remainder of 2; then 4 into 20 goes 5 giving an answer of 0.25

EXAMPLE 3

What is ⅛ ?

⅛ = 8) 1.000
or simply half of the previous .25. Think of the .25 as .250, thus half of 250 is 125, thus = .125
or ⅛ = ½ x ½ x ½
        = .5 x .5 x .5 (similar to 5^3)
5^3 is 125, but where does the Decimal Point go?
Because there are 3 digits and 3 decimal places, you need to move the decimal point 3 places to the right giving the right answer of .125

In this example, it is a good and useful practice to learn all the divisions of 1000, learning that an eighth of 1,000 is 125

EXAMPLE 4

What is 5/8 ?

5/8 = 5 x 1/8 = 5 x .125 = $.5/_10/_25$ = .625

EXAMPLE 4

What is 1/5 ?

1/5 = 5) 1.000   = .2

EXAMPLE 5

What is 1/25 ?

1/25 = 1/5 x 1/5 = .2 x.2 = .04

EXAMPLE 6

What is 1/10 ?

1/10 = 10) 1.000   = .1

EXAMPLE 7

What is 1/40 ?

1/40 = ¼ x 1/10 = .25 x .1 = .025

Calculate mentally, the following decimal forms of these non-recurring fractions:

a) **3/8 =**

b) **7/8 =**

c) **1/16 =**

d) **3/16 =**

e) **5/16 =**

f) **1/32 =**

g) **5/32 =**

Remove the decimal point from your answer and find the Square Root of this whole number. That is, what number squared, gives this number?

h) **28/32 =**

i) **1/64 =**

j) **3/64 =**

k) **1/128 =**

l) **2/5 =**

m) **1/20 =**

n) **1/80 =**

o) **1/100 =**

p) **1/400 =**

a) **3/8** = 3 x 1/8 = 3x.125 = .3/6/$_1$5 = .375

b) **7/8** = 7 x 1/8 = 7x.125 = .7/$_1$4/$_3$5 = .875

c) **1/16** = half of 1/8 = 2) .125 = .0625

d) **3/16** = 3x.0625 = .0/$_1$8/6/$_1$5 = .1875

e) **5/16** = 5x.0625 = .0/$_3$0/$_1$0/$_2$5 = .3125

f) **1/32** = half of .0625 = .03125

g) **5/32** = 5x.03125 = .0/$_1$5/5/$_1$0/$_2$5 = .15625
In terms of whole numbers .15625 = "15,625" since application of
the sutra: By One More   12 x 13 / 5^2 = 156/25
whose Square Root is 125 since 125^2 = 15,625

h) **28/32** = we need to simplify the numerator 28 and the
denominator 32 by dividing both numbers by 4 which is 7/8 which is
answer b) above = .875

i) **1/64** = half of .03125 = .015625

j) **3/64** =3x.015625 = .0/3/15/18/6/15 = .046875

k) **1/128** = half of 1/64 = half of .015625 = .0078125

l) **2/5** = 2x1/5 = 2x.2 = .4
   or double top and bottom = 4/10 = 4 x .1 = .4

m) **1/20** = ½ x 1/10 = .5 x .1 = .05

n) **1/80** = 1/8 x 1/10 = .125 x.1 = .0125

o) **1/100** = 1/10 x 1/10 = .1 x .1 = .01

p) **1/400** = ¼ x 1/100 = .25 x .01 = .0025

FULLY RECURRING
OR FULLY CIRCULATING DECIMALS

The denominators of a fraction that contain only 3, 7 or 11 or higher prime numbers as factors, will recur infinitely when there is not even a single factor of 2 or 5 in the denominator.

EXAMPLE 8

What is 1/3 ?

1/3 = 3) 1.000   = .333 forever.
We often place a dot over just one of the "3" and pronounce it as ".3 Repeater"
nb: do not confuse ".3 Repeater" with .3 which is the same as ".30000" which is very different, as ".30000" is the fraction of 3/10 and is less than ".3 Repeater" which is .333333 forever...

Using an Underline to express the same notation as the Dot that is meant to be placed above the repeating digits, ".3 Repeater" is thus represented as: ".3"
Here is the modern and correct way to express this:

$$\frac{1}{3} = .\dot{3}$$

EXAMPLE 9

What is 1/9 ?

1/9 = 9) 1.000 = .111 forever = ".1 Repeater"
= ".1"

or

$$\frac{1}{9} = .\dot{1}$$

EXAMPLE 10

What is 1/11 ?

1/11 = 11) 1.000   = .09 09 09 forever = "09 Repeater".
nb: in this case, there will be placed two dots, one over the "0" and one over the "9".
= ".09"

$$\frac{1}{11} = \cdot \overset{\cdot}{0}\overset{\cdot}{9}$$

EXAMPLE 11

What is 3/11 ?

3/11 = 3 x .09 = .27 Repeater or .2727272727272727
= ".27"

and check that 5/11 =

$$\frac{5}{11} = \cdot \overset{\cdot}{4}\overset{\cdot}{5}$$

EXAMPLE 12

What is 1/7 ?

1/7 = 7) 1.000   = .142857142857142857142857142857
= .142857 142857 142857 142857   or = .142857 Repeater
= .<u>142857</u> when the whole 6 repeating digits are dotted from above
= .<u>1</u>4285<u>7</u> repeater, but written more correctly like this:

$$\dot{\frac{1}{7}} = .\dot{1}4285\dot{7}$$

nb: these infinitely recurring 6 digits have a hidden pattern based on the Magic Of Nine. Look what happens when you place the last 3 digits underneath the first 3 digits:

1   4   2
8   5   7

Noticed how the 3 vertical columns all add up to 9.

There is a cryptic Vedic Sub-Sutra called:
"Kevalaih Saptakam Gunyat" which means:
"For 7 the Multiplicand is 143"
which gave the mathematician a reminder or a clue that to convert 1/7 into a decimal, he or she merely had to remember the number **"143"** subtract one from it, and that apply the sutra:
All from Nine And The Last From 10.
Thus 143 allowed the student to immediately arrive at the decimal form of

1/7 as 143 – 1 / 9-1/9-4/10-3 = 142 / 857

= .142857 142857 142857 repeater.

 All the digits between the 2 dots are also repeating, thus it has a Periodicity of 6.

EXAMPLE 13

What is 2/7 ? from the Wheel of 6 ?

Since this is a circulating decimal, it means that we can automatically determine what 2/7 is or what 3/7 is by sliding along the 6 digits contained in the 1/7 and determining the next lowest number. To determine 2/7 ignore the 1 of "142587" and find which is the next lowest number after the "1". It is not the "4" but the "2". We then read the numbers, beginning with the "2" going clockwise to get the order right.

Can you see that 2/7 = .285714 repeater. To understand this better, look at the hexagramic diagram below that depicts these infinitely recurring 6 digits as a cycle or wheel of 6ness.

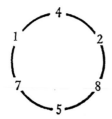

Starting from the Zenith, or top-most point, write the numbers 1-4-2-8-5-7 in a clockwise direction to assist you in calculating 3/7 and 4/7 and 5/7 and 6/7 asked of you in the following Exercise:

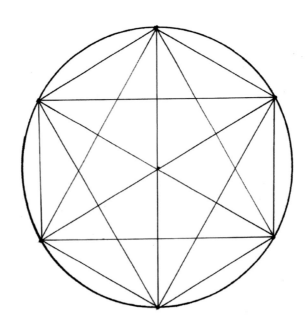

EXERCISE 2

Calculate mentally when possible, the following decimal forms of these fractions:

a) **2/3 =**

b) **2/9 =**

c) **7/9 =**

d) **2/11 =**

e) **5/11 =**

f) **8/11 =**

g) **10/11 =**

h) **3/7 =**

i) **4/7 =**

j) **5/7 =**

k) **6/7 =**

l) What happens when you multiply the decimal form of 1/7 by 7. ie: what is: .142857 x 7? Does this solution reveal another hidden pattern?

a) **2/3** = .333 or .3 repeater

b) **2/9** = 2x.1 = .222 or .2 repeater

c) **7/9** = 7x.1 = .777 or .7 repeater

e) **5/11** = 5x.09 = .45 45 45 or .45 repeater

f) **8/11** = 8x.09 = .72 72 72 or .72 repeater

h) **3/7** = .428571 428571 428571 or .428571 repeater

i) **4/7** = .571428 571428 571428 or .571428 repeater

j) **5/7** = .714285 714285 714285 or .714285 repeater

k) **6/7** = .857142 857142 857142 or .857142 repeater

l) **.142857 x 7 = .999999**
It appears in this Global Mathematics that the <u>Magic of Nine</u> is a hidden engine that runs smoothly to make all the patterns appear easily, as if it is some galactic mathematical base that all arithmetic is based upon.

AN UNUSUAL PROPERTY ABOUT THE RECIPROCAL OF 7

There exists a very strange property about the decimalized fraction of 1/7, ie: The Reciprocal of 7, as we have already seen, being .142857 Repeater.

Curiously, it involves the Binary Code or Doubling Sequence: (1 – 2 – 4 – 8 – 16 – 32 – 64 –) etc.

Let us multiply this mystical 7ness by the Doubling Sequence and observe the results we get on the RHS, but the results have been spread out in a peculiar way:

```
7 x 2¹ = 7 x 2   =     1 4
7 x 2² = 7 x 4   =       2 8
7 x 2³ = 7 x 8   =         5 6
7 x 2⁴ = 7 x 16  =         1 1 2
7 x 2⁵ = 7 x 32  =           2 2 4
7 x 2⁶ = 7 x 64  =             4 4 8
7 x 2⁷ = 7 x 128 =               8 9 6
7 x 2⁸ = 7 x 256 =               1 7 9 2
7 x 2⁹ = 7 x 512 =                 3 5 8 4
```
_____

                 1 4 2 8 5 7 1 4 2 8 5 7 1 4 2 (7 8 4)

nb: the results on the RHS have been arranged in a staircased manner such that the increasing steps are a length of 2 digits long, so that when we finally add all the sums, they display the curious repetition of 1-4-2-8-5-7.

nb: the final digits in brackets (7-8-4) are in error as they will be naturally affected by the progressive or next stages in the additions if we were to continue this madness. But it all works out quite beautifully!

(credit is given to Shakuntala Devi's classic book on Rapid Mental Calculation called "Figuring: The Joy Of Numbers" page 25)

This type of mathematical revelation or enquiry belongs to a very special branch of Mathematics known as **NUMBER THEORY**. This page, although appropriately inserted in the page on Fractions, really belongs in the next chapter on The Art of Number that includes some Number Theory.

UNICURSAL HEXAGRAM PUZZLE

Here is a puzzle:

This circle has 6 dots around its circumference, the grid or template for the construction of a typical Star of David. Can you draw another version of this star, called the UNICURSAL HEXAGRAM, by placing your pen or pencil upon any one dot, say the top dot, and continue to draw straight lines, without taking your pen off the paper, ensuring that all 6 dots are visited once, the goal being to return to the same top dot from where you started. The solution is elegant and surprisingly a beautiful 6-pointed Star Pattern rarely seen.
(It was a symbol used by Aleister Crowley, perhaps one reason why it is not commonly seen, as his work with magic was considered "dark").
I personally think it is a stark & beautiful pattern, and is encouraged to be used creatively.
Remember that it is drawn in a similar way to the construction of the typical Pentacle or 5-pointed star that we are all familiar with, starting at the top dot, of the 5 given dots, and straight lines drawn until you return to the same top dot, all 5 dots being connected.
Here is the grid:

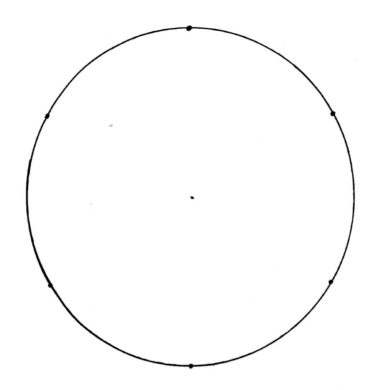

Here is the famous Pentacle an example of a Unicursal symbol, in the heart of the Paypaya fruit, and shown as a Magic Pentacle whose sums are 24.
The Magic Pentacle is drawn simply without taking the pen off the paper.

(Saravanan Thangaraja's daughter, Kuala Lumpur, 2005)

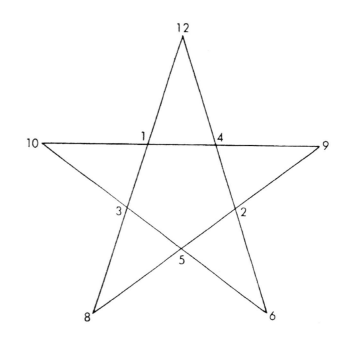

Any 4 numbers in a straight line have sums of 24.

Here are a 2 more empty circles to practice the Unicursal Hexagram:

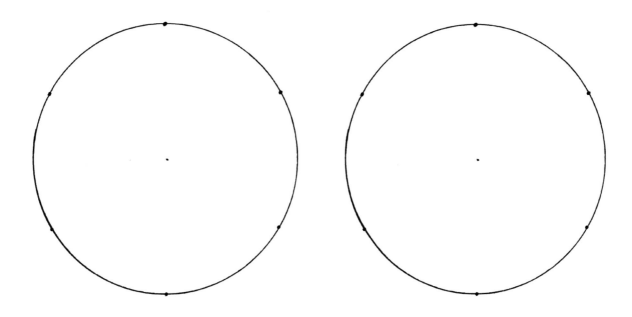

ANSWER 3

The UniCursal Hexagram.

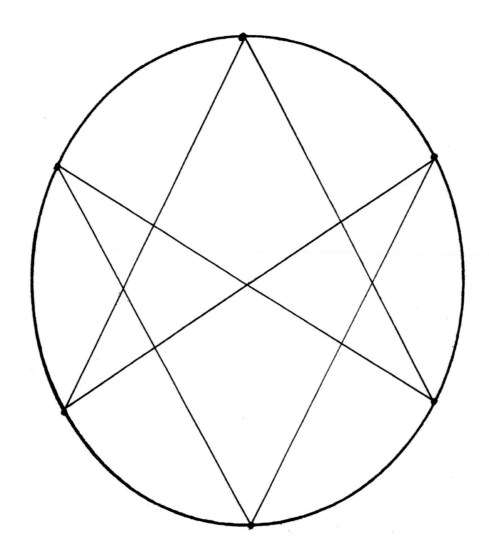

FRACTIONS

QUESTION:

What value is half way between 2.5 and 2.55 ?

ANSWER:

The half way point would be the Average, or Mean, which is the sum of both measurements, divided by 2.

```
2 . 5 0 +
2 . 5 5
_____
5 . 0 5
```

Now 5.05 divided by 2 is not an easy division, as there are some remainders to be dealt with, so best to add on another zeroe after the number 5.05

```
=  2 . 5 2 5
   _____
2 ) 5 . ¹0 5 ¹0
```

∴ the answer is 2.525

$$= 2.525$$
$$2 \overline{) 5 . {}^{1}0\ 5\ {}^{1}0}$$

EGYPTIAN FRACTIONS

The Egyptians always expressed their fractions with a numerator (top number) as a 1.

EXAMPLE 1

$$\frac{5}{6} = \frac{1}{2} + \frac{1}{3}$$

EXERCISE 1

Express the following fractions as Egyptian Fractions:

a) $\frac{3}{4}$ =

b) $\frac{7}{12}$ =

c) $\frac{8}{15}$ =

d) $\frac{9}{20}$ =

e) $\frac{9}{14}$ =

Answers: a) ¼ + ½   b) 1/3 + ¼   c) 1/3 + 1/5   d) ¼ + 1/5   e) ½ + 1/7

DAY 14

THE ART OF NUMBER

~ **Proof that 1=2**

~ **Difference of Two Squares**

~ **Number Theory** or **Art Of Number**
- **Odd Numbers Sequence** Generating the Squared Numbers
- Odd Numbers Sequence Generating the Cubic Numbers
- Pattern Of The 5th Powers Generated from the Odd Numbers
- Pattern Of The 6th Powers Generated from the Odd Numbers

PROOF of the ALGEBRRAIC FALLACY that 2=1

This proof shudders the whole foundation of modern world Mathematics.

Using only pure algebraic principles that every teenager student uses, as in: whatever you do to one side of the equation, you must do it to the other side, as if to maintain a balance, we do what we know, what we have been taught, yet lo and behold, 2=1.

For this fallacy to be demonstrated, it is important that the student has a quick review or memory-refresher on the topic known as: difference of two squares as a mental shortcut for subtracting say: $9^2 - 7^2$.

The important formula is written as:
DIFFERENCE OF TWO SQUARES

$A^2 - B^2 = (A-B).(A+B)$

where The Dot or " . " represents Multiplication "x".

EXAMPLE 1

$$15^2 - 13^2$$

Mentally calculate the difference of these 2 Squared Numbers:
$15^2 - 13^2$
Let the two variables be assigned their values as:
a=15 and b=13
These numbers need to be substituted or plugged into the following equation:

$a^2 - b^2 = (a-b).(a+b)$

$$15^2 - 13^2 \quad = (15 - 13). (15 + 13)$$
$$= 2 \times 28$$
$$= 56$$

Now, that you have remembered the Difference Of Two Squares, you will be able to intelligently apply it to the following PROOF to reveal the fallacy that **2=1**.

PROOF of the ALGEBRRAIC FALLACY that 2=1

Let the 2 variables have the same value

$a = b$

Multiply both sides by a

$a^2 = ab$

Subtract b^2 from both sides

$a^2 - b^2 = ab - b^2$

Expand the LHS
(this is where you need to substitute the expression for the difference of two squares) and simplify the RHS

$(a - b).(a + b) = b(a - b)$

Divide both sides by $(a - b)$

$$\frac{(a - b).(a + b)}{(a - b)} = \frac{(a - b)b}{(a - b)}$$

cancelling out $(a - b)$ from top and bottom

$(a + b) = b$

but a = b so substitute this value into the equation

$2b = b$ which is really $2b = 1b$ then divide both sides by b

2 = 1

???

What, 2 = 1
how can this be?

How do we explain this fallacy.

A professor might respond to this, to salvage his or her degree, and say that the error was in assuming that: "If a & b are 2 variables, they are not allowed to be equal to one another"

So to test this argument, to refute this argument, is to try to find an example in high school algebra, where a & b, the two variables, are equal. Here is a clear example that shows that a can equal b and thus a & b can be equal.

Look at the case for $(a+b)^2$ when it is expanded.
We know that $(a+b)^2 = a^2 + 2ab + b^2$
So let us plug a=2 and b=2 and see that the LHS is the same as the RHS, which is the case:
$(2+2)^2 = 2^2 + 2x2x2 + 2^2$
thus does $4^2 = 4 + 8 + 4$
does 16 = 16
Yes
Therefore 2 variables can have the same value
Yes, we can legally start with the premise that a=b
Yes, this is an established Law of Modern Algebra that permits variables to be of the same value

Yet 2=1 is still an Enigma!!!

There is something seriously wrong with the evolution of our mathematical genii that allows 2=1 using all the rules that we have inherited.
This is quite a serious argument, yet to be solved...

This anomaly is good though, since anomalies allow seekers to travel down the rabbit-hole and emerge with new insights, that change the world.

ODD NUMBERS
GENERATING THE SQUARED NUMBERS
UNIVERSAL LANGUAGE of PATTERN RECOGNITION

EXERCISE 1

Using your Pouch of 25 Pebbles, what is the common SHAPE of the Answers to the following Sums of the Odd Numbers?

Knowing this shape, mentally predict the answers, by filling in the box below to the questions c, d, e, f, g, h, i, j, k, l and m.

What is your Conclusion?

a.	1	= 1	= 1 x 1	= 1^2
b.	1 + 3	= 4	= 2 x 2	= 2^2
c.	1 + 3 + 5	=	=	=
d.	1 + 3 + 5 + 7	=	=	=
e.	1 + 3 + 5 + 7 + 9	=	=	=
f.	1 + 3+ 5 + 7 + 9 + 11	=	=	=
g.	1 + 3 + 5 + 7 + 9 + 11 + 13	=	=	=
i.	1 + 3 + 5 + 7 + 9 + 11 + 13 + 15	=	=	=
j.	1 + ... + 17	=	=	=
k.	1 + ... + 19	=	=	=
l.	1 + ... + 21	=	=	=
m.	1 + ... + 23	=	=	=

The concept being demonstrated here is that the **Supreme Language of Pattern Recognition** or **Universal Shape**, is operating.

The new mathematical education or revolution is the ability to see numbers as pictures or shapes.

It is the **Right Feminine Brain Mathematics** that understands pictures, shapes, music, holograms, patterns, symbols etc.

By learning this language, we are remembering the lost science of the Human–Dolphin Brain Connection. Research shows that Dolphins have the ability to do 8 things at once, like listening to 8 radio channels simultaneously! reflecting that we do have this ability.

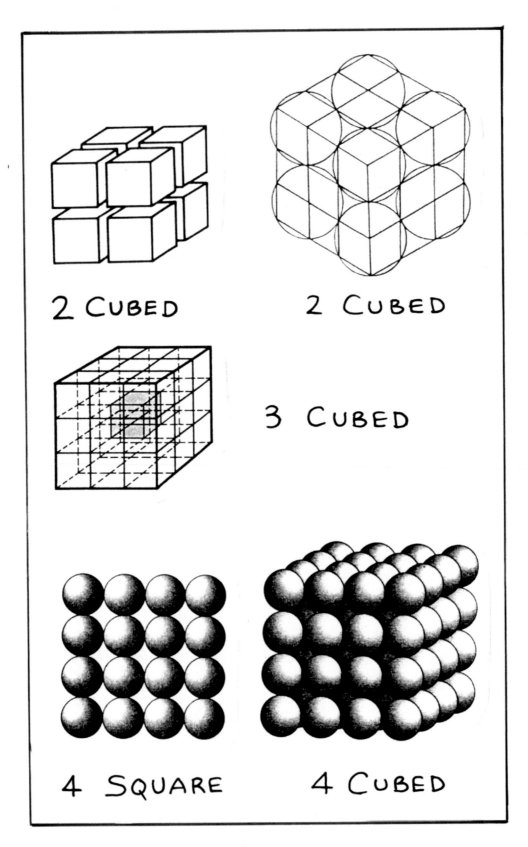

2 CUBED

2 CUBED

3 CUBED

4 SQUARE

4 CUBED

Learning to view **Numbers as Shapes** is the first step to grokking the Galactic Mathemagics known here as THE ART OF NUMBER, or the Translation of Number Into Art.

ODD NUMBERS
GENERATING CUBED or CUBIC NUMBERS

EXERCISE 2

In Exercise 1, we saw how the Odd Nos. generated the Squared Nos. To move **Into The Next Dimension**, (Jain's 17th Sutra) with this knowledge, is to know how to go from the Family of Squares into the Family of Cubes, and using only ODD NUMBERS !
Add these sums :

a.	1	= 1	= 1 x 1 x 1	= 1^3
b.	3 + 5	= 8	= 2 x 2 x 2	= 2^3
c.	7 + 9 + 11			
d.	13 + 15 + 17 + 19			
e.	21 + 23 + 25 + 27 + 29			
f.	31 + 33 + 35 + 37 + 39 + 41			
g.	43 + 45 + 47 + 49 + 51+ 53 + 55			

The Sutra that is at work here is the first One:
Known as 'BY ONE MORE THAN THE PREVIOUS DIGIT"

eg :

1	=	1 x 1
1 + 2 + 1	=	2 x 2
1 + 2 + 3 + 2 + 1	=	3 x 3
1 + 2 + 3 + 4 + 3 + 2 + 1	=	4 x 4

Noticing the stair-casing of digits going up (ascending), then going down (descending)!

ANSWERS 1

The Odd Numbers, added up in this pyramidal fashion generates the Sequence of Squared Numbers:

1 – 4 – 9 – 16 – 25 – 36 – 49 – 64 – 81 – 100 – 121 – 144 etc

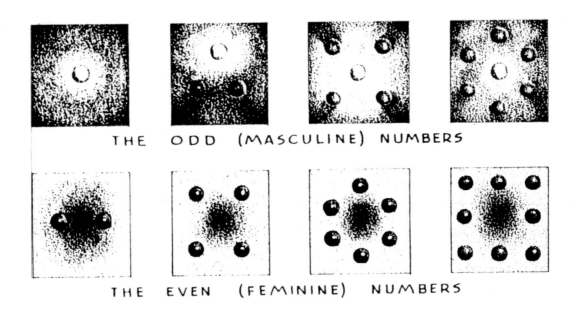

THE ODD (MASCULINE) NUMBERS

THE EVEN (FEMININE) NUMBERS

ANSWERS 2

The Odd Numbers, added up in this pyramidal fashion generates the Sequence of Cubed Numbers:

1 – 8 – 27 – 64 – 125 – 216 – 343 – 512 – 729 – 1000 – 1331 – 1728 etc

The ancient school of Pythagoras, 2,500 years ago, saw cosmological meaning in the partitioning of their reality into Odd (Masculine) and Even (Feminine) Numbers.

ODD NUMBERS SEQUENCE
GENERATING The 5th & 6th POWERS

Here are 2 more examples of Patterns hidden in the sequence of Odd Numbers:

If anything, these two charts are a good start for the student to begin learning the Sequences of The 5th Powers and 6th Powers.

REVELATION of the FIFTH POWERS (n^5)
and Introducing the Concept of "Skipping or Missing" Odd Numbers!

Amount of Nos Added	Sequence of Odd Numbers	Sum	Powers
1	1	1	1^5
4 but **Skip 1**	5+7+9+11	32	2^5
9 but **Skip 3**	19+21+23+25+27+ 29+31+33+35	243	3^5
16 but **Skip 6**	49+51+53+55+57+59+61+ 63+65+67+69+71+73+75+ 77+79	1024	4^5

Pattern Of The 5th Powers Generated from the Odd Numbers (but has some Odd Numbers missing!).

REVELATION of the SIXTH POWERS (n^6)

Amount of Nos Added	Sequence of Odd Numbers	Sum	Powers
1 or 1^3	1	1	1^6
8 or 2^3	1+3+5+7+ 9+11+13+15	64	2^6
27 or 3^3	1+3+5+7+9+11+13+15+17+19+21+23+25+27+29+31+33+35+37+39+41+43+45+47+49+51+53	729	3^6
64 or 4^3	1+3+5+7+9+11+13+15+17+19+21+23+25+27+29+31+ ....+127	4096	4^6

Pattern Of The 6th Powers Generated from the Odd Numbers

DAY 15

SQUARING & SQUARE ROOTS

~ SQUARING OF NUMBERS ENDING IN 25 eg 125^2

~ SQUARE ROOTS $\sqrt{}$
- 6 Step Method How to Calculate Square Roots that contain either 3 or 4 Digits eg: $\sqrt{729}$ and $\sqrt{4906}$
- Table of End Digits of a Square Root
- Square Roots of Numbers from 10,000 to 40,000 Utilizing the Table of Teen Numbers Squared.

~ GRAPHICAL MULTIPLICATION
- Method 1: The Stroke Method eg: 12 x 35

SQUARING NUMBERS ENDING IN 25

EXAMPLE 1

$$125^2$$

125^2 = has 3 parts: ___ / ___ / ___

= square the left digit / product of outer digits / square the 25
= 1^2 / 1x5 / 25^2
= 1 / 5 / 625
= 15,625

nb: all numbers ending in 25 when squared have their last 3 digits as 625.

325^2 = 3^2 / 3x5 / 25^2
= 9 / $_1$5 / 625
= 10 / 5 /625
= 105,625

EXERCISE 1

Mentally Square the following 3 digit numbers ending in 25 and then show the working out as a one-line answer, in the style shown above.

a) 225^2 =

b) 425^2 =

c) 525^2 =

d) 625^2 =

e) 725^2 =

f) 825^2 =

g) 925^2 =

ANSWERS 1

a) 225^2 = 2^2 / 2x5 / 625 = 4 / $_1$0 / 625   = 5/0/625   = 50,625

b) 425^2 = 4^2 / 4x5 / 625 = 16 / $_2$0 / 625 = 18/0/625   = 180,625

c) 525^2 = 5^2 / 5x5 / 625 = 25 / $_2$5 / 625 = 27/5/625   = 275,625

d) 625^2 = 6^2 / 6x5 / 625 = 36 / $_3$0 / 625 = 39/0/625   = 390,625

e) 725^2 = 7^2 / 7x5 / 625 = 49 / $_3$5 / 625 = 52/5/625   = 525,625

f) 825^2 = 8^2 / 8x5 / 625 = 64 / $_4$0 / 625 = 68/0/625   = 680,625

g) 925^2 = 9^2 / 9x5 / 625 = 81 / $_4$5 / 625 = 85/5/625   = 855,625

SQUARE ROOTS
(of Numbers under 10,000)

$\sqrt{}$

Let "x" be the root of a number, that when it is multiplied by itself, it results in "x". ("x" here does not mean the symbol for multiplication, it is the 24th letter of the English Alphabet).

The symbol for this Square Root is "√" & is called the Radical symbol.

Thus, algebraically, it is written like this:

(√x) times (√x) = x   as in: 8 times 8 = 64

Look at Table 1 below and observe that "8 Squared" = 64, thus the Square Root of 64 (√64) = 8

Table 1

Number	Square
1	1
2	4
3	9
4	16
5	25
6	36
7	49
8	64
9	81
10	100

Table 2

End Digit Of a Square	End Digit of a Square Root
1	1 or 9
4	2 or 8
9	3 or 7
6	4 or 6
5	5
0	0

nb: in Table 2, observe on the RHS, The End Digits of a Square Root form Pairs of 10. Knowing this will assist you in determining Square Roots of numbers under 10,000.

It is also easier to determine Cube Roots than it is Square Roots, for this reason, that Square Roots involve the memorization of these Pairs; it is these Pairs of 10 that will create the Possibility of two Answers, and your job or task is to know how to discern between the two Possible Answers, this will be explained soon as it involves the concept of "Averages".

Also, in this brief chapter, I am only interested in whole numbers or integers, no decimalized numbers, only Perfectly Squared Numbers. So we will not look at say $(\sqrt{19})$ =4.356... as it is not an integer.

Just for your knowledge base, The Square Root of a number, on a computer screen is written often as a Fractional Exponent, meaning if we want to express the Square Root of 64 ($\sqrt{64}$) = 8 it is written as **$64^{1/2}$** = 8 meaning: "64 raised to the Half Power of Unity equals the square root of that number = 8".

Let us examine 2 examples, Example 1 is in detail, then in Example 2 the steps are simplified so that you can use these simplified methods to calculate the square roots in the Exercises given at the end of this chapter.

6 STEP METHOD HOW To CALCULATE SQUARE ROOTS That CONTAIN EITHER 3 or 4 DIGITS:

EXAMPLE 1

What is √4096 ?

What is the Square Root of 4,096?

1. Partition the given number into 2 parts, starting from the RHS, counting two steps or digits, and insert a "/" or "Forward Slash" symbol:

40 **/** 96

2. For each of the 2 Partitions or Sections, one being the LHS and the other being the RHS, we draw a horizontal bar underneath, to indicate that it will be a 2 Digit Answer:

40 **/** 96

__ / __

3. Start with the LHS Pair "40" you scroll down the numbers in Table 1, looking for the first square number below 40.
It is 6 since $6^2 = 36$.
Or you could think of "40" being sandwiched between 2 Squares, between 6^2 and 7^2, (an important concept, as soon you will need to be multiplying these two squares to determine an Average).
Here is the setting out so far:

40 **/** 96

_6_ / __

Continued...

4. Here now is the tricky bit, that makes Squares Roots harder than Cube Roots. We need to calculate the digit, from 1 to 9, for the RHS or The Units Column. Here we have to choose from 2 Options or 2 Possibilities. We know that 409**6** ends in a 6, so we scroll down in Table 2 and look for the End Digit that ends in a 6 and learn that it can either be a 4 or a 6. This gives us our **2 Options**, for the final answer to **64 or 66**.

5. There is a Test of Averages to determine the correct option or possibility for the end digit of the Two Possible Square Roots.
We noticed before that the LHS Pair "40" (of **40**96) was sandwiched between the two squares: between 6^2 and 7^2, ie: between 36 and 49. Now the Average or Mean of these two numbers is 6x(6+1) or 6x7 =42

6. Here is the last part: Since the Average is 42, we have to establish whether or not this number of 42 is Less Than (<) or Greater Than (>) the LHS Pair number "40"?
Since 42 is > 40, we choose the smaller option of the 2 numbers, which is 64, (rather than 66) and have our Answer:

40 **/** 96

 6 / 4

Thus $\sqrt{4096}$ = **64**

EXAMPLE 2

What is $\sqrt{7744}$?

(Use this 6-Step Method of setting out for the upcoming Exercises).

Find $\sqrt{7744}$ = ?

1. 77 / 44

2. __ / __

3. First Digit on LHS = 8 (since 77 is in between 8^2 and 9^2)
   = _8_ / __

4. Last Digit = 2 or 8 (from Table 2, because the Square ends in a 4).
   This means that the 2 Possible Roots are therefore 82 or 88

5. Since 77 lies between the 2 Squares of 8^2 and 9^2, then the Average is 8x9 = 72

6. Since 72 is < 77, we select the larger of the 2 numbers seen in Step 5 and arrive at our Answer of = _8_ / _8_

Thus $\sqrt{7744}$ = 88

Find the Square Roots ($\sqrt{\ }$) of the following
Perfect Squared Numbers.
If possible, do them Mentally as One-Line Answers.

a) $\sqrt{729}$ =

b) $\sqrt{1,089}$ =

c) $\sqrt{1,764}$ =

d) $\sqrt{2,916}$ =

e) $\sqrt{4,761}$ =

f) $\sqrt{5,776}$ =

g) $\sqrt{6,561}$ =

h) $\sqrt{9,604}$ =

a) $\sqrt{729}$ = 7/29 = 2 / _ = 23 or 27. Average = 2x3=6
which is <7 ∴ choose the greater option = **27**

b) $\sqrt{1,089}$ = 10/89 = 3 / _ = 33 or 27. Average = 3x4=12
which is >10 ∴ choose the lesser option = **33**

c) $\sqrt{1,764}$ = 17/64 = 4 / _ = 42 or 48. Average = 4x5=20
which is >17 ∴ choose the lesser option = **42**

d) $\sqrt{2,916}$ = 29/16 = 5 / _ = 54 or 56. Average = 5x6=30
which is >29 ∴ choose the lesser option = **54**

e) $\sqrt{4,761}$ = 47/61 = 6 / _ = 61 or 69. Average = 6x7=42
which is <47 ∴ choose the greater option = **69**

f) $\sqrt{5,776}$ = 57/76 = 7 / _ = 74 or 76. Average = 7x8=56
which is <57 ∴ choose the greater option = **76**

g) $\sqrt{6,561}$ = 65/61 = 8 / _ = 81 or 89. Average = 8x9=72
which is >65 ∴ choose the lesser option = **81**

h) $\sqrt{9,604}$ = 96/04 = 9 / _ = 92 or 98. Average = 9x10=90
which is <96 ∴ choose the greater option = **98**

SQUARE ROOTS OF NUMBERS UNDER 40,000

So far, we can calculate the square roots of numbers up to 10,000. If we wanted to expand our possibilities, say up to 40,000, we would have to learn the sequence of Squared Numbers in the Teens, up to 19^2.
(With a bit more knowledge, you can calculate the Square Roots of a Number with 5 or 6 digits, which would have a root of 3 digits, as in the $\sqrt{998,001}$ which is 999).

EXAMPLE 3

What is the $\sqrt{20,736}$?

Table 3: **The Teen Numbers Squared**

Number	Square
11	121
12	144
13	169
14	196
15	225
16	256
17	289
18	324
19	361
20	400

(Use this 6-Step Method of Setting Out).

Find $\sqrt{20,736}$ = ?

1. 207 / 36

2. __ / __

3. First Digit on LHS = 14 (since 207 is in between 14^2 and 15^2)
 = _14_ / __

4. Last Digit = 4 or 6 (from Table 2, because the Square ends in a 6).
This means that the 2 Possible Roots are therefore 144 or 146

5. Since 207 lies between the 2 Squares of 14^2 and 15^2, then the Average is 14x15 = 210

6. Since 210 is > 207, we select the lesser of the 2 numbers seen in Step 5 and arrive at our Answer of __14__ / __4__
Thus $\sqrt{20,736} = 144$

EXERCISE 4

Find the Square Roots ($\sqrt{}$) of the following
Perfect Squared Numbers that have 3 Digits in their Root.

a) $\sqrt{15,129}$ =

b) $\sqrt{18,496}$ =

c) $\sqrt{23,409}$ =

a) $\sqrt{15,129}$ = 151/29 = 12 / _ = 123 or 127. Average =
12x13=156   which is >151   ∴ choose the lesser option = **123**

b) $\sqrt{18,496}$ = 184/96 = 13 / _ = 134 or 136. Average =
13x14=182   which is <184   ∴ choose the greater option = **136**

c) $\sqrt{23,409}$ = 234/09 = 15 / _ = 153 or 157. Average =
15x16=240   which is >234   ∴ choose the lesser option = **153**

nb: These 3 answers: 123 and 136 and 153 have special qualities.
Do some research and identify what these properties are.

eg: **$153 = 1^3 + 5^3 + 3^3$**
Such analysis of inherent patterning is called <u>NUMBER THEORY</u>.

For further knowledge of how to perform square roots of numbers having 5 or 6
or digits, like the $\sqrt{186,624}$, consult the following 2 books.

Mathematical genius
Shakuntala Devi is an Indian woman of immense arithmetical ability who demonstrates her amazing skills in public. It took just a second for her to find that the cube root of 332,812,557 is 693. In the 1970s she calculated the 23rd root of a 201-digit number. The problem had been devised by Texan students. A computer took one minute and 13,000 instructions to check her answer and prove her right.

"All along I have cherished a desire to show those who think mathematics boring and dull just how beautiful it can be. This book is what I offer."
Shakuntala Devi

Shakuntala Devi

Figuring The Joy of Numbers

1. Shakuntala Devi, authoress of "**The Joy Of Numbers**" is an Indian woman of immense arithmetical ability who demonstrates her amazing skills in public. It took just a second for her to find that the cube root 333,812,557 is 693. In the 1970s she calculated the 23rd root of a 201-digit number. The problem had been devised by Texan students. A computer took one minute and 13,000 instructions to check her answer and prove her right.

2. Kranthi Kiran Tumma known as a Walking Genius, in his early years, traveled extensively teaching at many institutions of higher learning, the supreme art of Rapid mental Calculation. His best known book is called:
"**Calculations @ The Speed Of Light.**
When I met him in India in 2005, he personally assisted me in teaching to many large audiences seeking to learn Vedic Mathematics, before it became popular.
He is shown here with his wife Radhika.

GRAPHICAL MULTIPLICATION

This section will outline two brilliant methods of how to perform basic multiplication, without having to know your traditional multiplication times table, nor having to use the concept of numbers, which means, you are about to do **mathematics via the supreme language of pictures or shapes.**

Please take time to understand what is being presented, as this material is truly astonishing, you will be abandoning your left logical rational male cortexial side of your brain (which deals with the world of numbers), and you will gracefully switch over to the visual, intuitive, female cortexial side of your brain (which deals with shapes and patterns).

This is an important part of Jain Mathemagics, as the simple principle of: THE TRANSLATION OF NUMBER INTO ART does apply here, and is a critical part in the future global mathematical curriculums.

Though we would still prefer to do our mathematical calculations in our head, that is, by rapid mental calculation, the following method is still valid to learn and know, as it will introduce to the student valuable ideas of Number Theory. (nb: there are other methods of Graphical Multiplication to be taught, eg: "The Ring Method" not published in this book, but available for the serious student).

METHOD 1:
The STROKE METHOD

Let us begin by multiplying **12 x 35**.

We have deliberately chosen the smaller numbers to make the example easier to explain.

We begin by drawing a large square that will contain some intersecting lines. The first part of the calculation, the "12" of the "12x35", will be represented by vertical lines, and the second part of the calculation, the "35" of the "12x35", will be represented by horizontal lines.

To understand the following description of how to visually determine the answer to 12 x 35, I will insert the solution, Fig A below, so that you can "**see**" the answer!

GRAPHICAL MULTIPLICATION

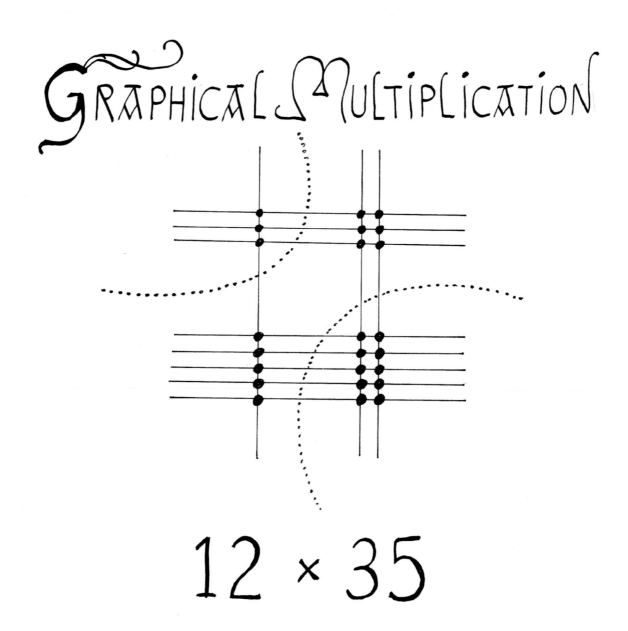

12 × 35

Fig A
The "Stroke Method"
for calculating 12 x 35

Can you see how the "**12**" of the "12x35", is represented by the above vertical lines, but notice that there is one vertical line separated by two vertical lines, being the 1 and the 2 of the "12" respectively.

Now we are going to cross this data of vertical lines, with some horizontal lines representing the "**35**" of the "12x35".

Can you see how the "35" of the "12x35", is represented by the above horizontal lines, but notice that there are 3 horizontal lines separated by five horizontal lines, being the 3 and the 5 of the "35" respectively.

Notice how I have highlighted the intersection points or the 4 nodes or nexii (plural of nexus or meeting points) by placing thick circular dots over these nexii.
From the top to the bottom, the 4 nexii can be counted as:
3 dots, 6 dots, 5 dots and 10 dots.

The answer is staring us in the eyes, but the 4 nexii have to be viewed as 3 partitions or segments to grok it.
I have drawn, in Fig A, some curved dots, as this will help you to see how the quadrature is trinitized (or how the 4 nexii are divided into 3 parts).
These 3 parts, viewed from the Left to the Right, are really the Place Value System of Decimal Mathematics, (the Hundreds, Tens and Units columns) meaning that:

•   The far left hand side (which contains the 3 dots) is the hundreds column, thus representing 300.
•   The middle section between the two curves must be combined or added together, which means we inspect the diagonal nature of this middleness (which is the sum of 6 dots and 5 dots), making 11 dots in total, and ascribe to this number, the place value system of the 10s Column.
•   The far right hand side (which contains the 10 dots) is the Units Column, thus representing the number 10.

We must now combine these 3 segments into a single line of arithmetic, but it will be read from Right to Left, which is the ancient Indian or Vedic method of reading mathematical data, opposite to what we have learnt here in the West.

Thus 12 x 35
= 3 / 6+5 / 10 (being suggestive of the Hundreds/Tens/Units).
= 3 / 11 / 10
= 3 / $_1$1 / $_1$0

Notice how the "11" is written as "$_1$1" with the subscript, and the "10" also is written as "$_1$0" with the subscript. This is because there is a Indian mathematical rule that only one digit is allowed to occupy the Units and Tens Column, in this case, and that these subscripted numbers will be used as carry-overs to the left. Starting with the "$_1$0" we are going to carry-over the subscripted "1" to the number "11" on the left, which makes "12" but will be written as "$_1$2".

= 3 / $_1$2 / 0

And we do the same again, we carry over the subscripted 1 of the "$_1$2" over to the left to be added to the 3 which gives:

= 4 / 2 / 0

= 420

(Just a note to explain the "Forward Stroke" or "/" used in the above examples. It was introduced by the late Pope and Scholar of India: Bharati Krsna Tirthaji (1884 – 1960). It acts as a "place marker" like in our place value systems of Hundreds/ Tens/ Units Columns, without it, there is more likely a chance for an error to occur. Basically, if Vedic Mathematics, aka Rapid Mental Calculation gets adopted by the western world in the near future, it means we will have to adopt this "/" which is quick clever and useful when you become acquainted to other methods in how it can be used).

(There is a .wm file reference
called "Multiplication With Strokes"
that was sent to me from an email from Chezoslavakia:
(gyanpuri@tiscali.cz wrote: Hello Jain, see attachment. Gyanpuri)
a must see, to understand how fluid and easy it is to multiply in this fashion.
Since this is a highly visual experience, it is recommended that you get to view this file, if it is still available. I have referenced it here only to portray to the reader how this video file was filmed, it was done with no words, only a pair of silent hands forming those relevant strokes. Quite Brilliant when viewed for the first time.

There is also another you tube clip
that shows the 2nd Method aka The Ring Method
and it is called: Another Graphical Multiplication Trick
and the link is:
http://www.youtube.com/watch?v=0GlIx5pztpM

It shows only a hand with a pen being held, and writing upon a piece of paper, the symbols of the rings being divided by the numbers to arrive at the correct answer).

One of the shortfalls or negatives about this Stroke Method is that it becomes a bit complex when using numbers larger than 5 to multiply. eg: to multiply: 67x89 would be time consuming, but it would work. So best to apply this method when using digits under 5 like those shown in the Exercise below.

EXERCISE 2

Use this Stroke Method to calculate the following multiplications:

a) 12 x 34   =

b) 23 x 54   =

c) 13 x 46   =

d) 123 x 321 =

e) 18 x 18   =

f) 108 x 108   =

Here are the answers for **a)**, **b)**, **c)** and **d)**

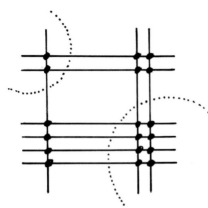

12 × 34

$= 3 / 10 / 8$

$= 3 /_1 0 / 8$

$= 4 / 0 / 8$

$= 4 \ 0 \ 8$

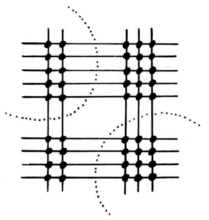

23 × 54

$= 10 / 23 / 12$

$= 10 /_2 3 /_1 2$

$= 12 / 4 / 2$

$= 1,2 \ 4 \ 2$

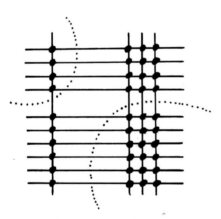

13 × 46

$= 4 /_1 8 /_1 8$

$= 5 / 9 / 8$

$= 5 \ 9 \ 8$

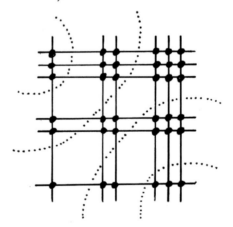

123 × 321

$= 3 / 8 /_1 4 / 8 / 3$

$= 3 / 9 / 4 / 8 / 3$

$= 3 \ 9 , 4 \ 8 \ 3$

Here is the answer for **e) 18 x 18**

18 × 18

$$= 1/16/64$$

$$= 1/{}_16/{}_64$$

$$= 1/{}_22/4$$

$$= 3/2/4$$

$$= 324$$

(nb: Base 10 allows only 1 digit after a "forward slash").

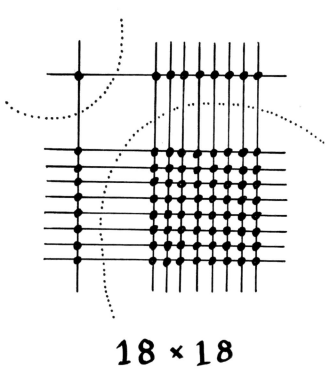

18 × 18

Here is the answer for **f) 108 x 108**

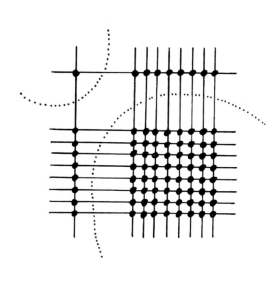

108 × 108

$$= 1/16/64 \qquad = 11,664$$

nb: Base 100 allows only 2 digits after a "forward slash".

∴ In this example, there are no "carry-overs".

DAY 16
CUBIC ROOTS:

$$\sqrt[3]{}$$

The Sutra that applies here is "Vilokanam"
aka "By Mere Observation"
aka "By Mere Intuition"

- Revision of the Previous Week.
- Revise Trachtenberg Method that gives the Cubic
  Powers eg: 8x8x8 = 64x8 = 512
- Learning the Chart of The Sequence of Cubic Numbers
- Learning the Final Digits of the Cubic Numbers and how they relate
  to the Complements or Pairs of 10
- Using "Digit Sums" to determine if a number is a Perfect Cube

TRACHTENBERG METHOD
TO SOLVE THE SEQUENCE OF CUBIC NUMBERS

At this stage, we are going to do another set of exercises, the same again with 2D x 1D (short for 2 Digits x 1 Digit) to determine the Sequence of CUBIC NUMBERS. It will be necessary to remember these Cubic Numbers, as you will be required to learn how to mentally solve Cubic Roots achieved by mentally remembering this Sequence of Cubic Numbers:

1, 8, 27, 64, 125, 216, 343, 512, 729 ...

EXAMPLE 6x6x6

6x6x6 is abbreviated to "**6 Cubed**" or "**6 To The Power of Three**" or "**6^3**". This Power of 3 is also known as an "**Index**" whose plural form is "**Indices**". Another synonym or word that means the same thing is: "**Exponent**".

To rewrite 6x6x6 in the Trachtenberg form of 2D x 1D we write the 6x6 part as 36 and multiply this by 6.

Thus 6x6x6 = 36x6

Apply the principle of multiplying the 36 by 6 expressed as two pairs:
36x6 = 18 / 36 but the rule is we are only allowed one digit on the right hand side of the forward slash, so we express 6x6x6 as:

18 / $_3$6 then slide the "baby 3" or "carry-over" to the left hand side, giving the final answer as 21/6 or **216**.

As a Mental, One-Line Answer "**6^3**" can be written in this format:
6^3 = 6x6x6 = 36x6 = 18 / $_3$6 = 21/6 = **216**.

Give the answers to the following Cubic Numbers mentally, then write out the One-Line Answer format as shown above:

a) $4^3 =$

b) $5^3 =$

c) $7^3 =$

d) $8^3 =$

d) $9^3 =$

nb: In this exercise, you will see that you will have to know your Times Tables, testing your knowledge of "Magic Fingers" to calculate the Series of Squared Numbers like 7^2 aka 7x7.

ANSWERS to Exercise 2

a) 4^3 = 4x4x4 = 16x4 = 4 / $_2$4 = 6/4 = **64**.

b) 5^3 = 5x5x5 = 25x5 = 10 / $_2$5 = 12/5 = **125**.

c) 7^3 = 7x7x7 = 49x7 = 28 / $_6$3 = 34/3 = **343**.

d) 8^3 = 8x8x8 = 64x8 = 48 / $_3$2 = 51/2 = **512**.

e) 9^3 = 9x9x9 = 81x9 = 72 / 9 = **729**.

CUBE ROOTS or CUBIC ROOTS

This is an amazing trick which has astounded the audiences I have addressed in various workshops around the world. This jaw-dropping and awe inspiring technique helps you find out the cube root of a 4 or 5 or 6 digits number mentally.

Here are some notes to get started:

1) Cube of a 2-digit number will have at max 6 digits (99^3 = 970,299). This implies that if you are given a 6 digit number, its cube root will have 2 digits.

2) This trick works only for perfect cubes, it will not work for any arbitrary 6-digit number.

3) It works only for whole number integers

4) After some practice, you can do cube roots in less than a second, that is, literally at the speed of thought!

This is an excellent **Party Trick** to determine the Cube Root of a 5 or 6 digit number using Rapid Mental Calculation.

So our Question is:

WHAT IS THE CUBE ROOT OF **262,144**?

| EXAMPLE 1 |

$$\sqrt[3]{262,144}$$

It is known that it's a perfect cube.

It is written correctly as shown above, where the numbers are written under a long horizontal line called the "Radical" or "Surd" symbol but this symbol is lacking when written in computer language, so we need to express the Inverse of the 3rd Power of n^3 in its fractional form as $n^{1/3}$ to represent the Cube Root, and create a symbol "^" to express Powers also known as Exponents or Indices. We therefore write the question like this:

$(262,144)$^1/3 = ?

It simply means, 262,144 raised to the Power of One Third, or what is the cube root of 262,144 or what number multiplied by itself 3 times gives this number of 262,144?

[If we wanted to write in computer language, What is the Square root of 64? we would write it as (64)^1/2 = 8]

Now divide this number 262,144 into two parts. Starting from the far right hand side, draw a line so that there are distinctly 3 digits. The remaining digits will be isolated on the left hand side. This is what it will look like, shown below.

262   /   144

----- / -----

You know the answer will have 2 digits. There will be a Digit in the Tens Column and a Digit at Unit's Column. You will determine the digit in the Ten's Column by using the left hand side of the original number (262) and digit in the Unit's Column by using the right hand side of the number (144)

Step 1.

It is important that you memorize the following Cubic Numbers, known as The Sequence of Cubes.

Chart 1: Cubic Numbers from 1 to 10

Number	Cube
1	1
2	8
3	27
4	64
5	125
6	216
7	343
8	512
9	729
10	1000

Chart 2: Unit's digit of the Cube Roots

Cube Ends in	Cube Root Ends in
1	1
2	8
3	7
4	4
5	5
6	6
7	3
8	2
9	9
0	0

With a pencil, draw curved lines in Chart 2, on the RHS column, connecting the numbers that have a sum of 10. These Pairs or Complements of 10, when learnt to memory, will assist you in doing cubic roots at the speed of light.

Step 2.

For the left hand side we need to use Chart 1. We have to scroll down the list of memorized cubic numbers to establish quickly which cubic number is less than our number of 262. In this case we see that our number of 262 is above the perfect cube of 6 or 216, so we select the number "6" as the Tens Digit of our answer.

We therefore write the "6" into its correct placement:

262   /   144

  6

-----   /   -----

So far we have half our answer.
The answer is "60 something"!

Step 3.

For right hand side we need to use Chart 2 to establish the Unit's Digit. Since our original number (the perfect cube) ends in 4 (see 262,14**4**). Its cube root will end in 4.

Thus the units digit will be 4. We write this in its correct placement:

262   /   144

  6   **4**

-----   /   -----

Combining the results we get the answer as 6/4 which is really 64.

Thus $(262{,}144)^{1/3} = 64$

| EXAMPLE 2 |

WHAT IS THE CUBE ROOT OF **35,937**?

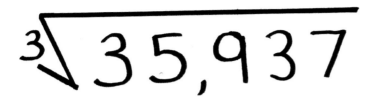

$3\sqrt{35{,}937}$

35 / **937**

3   **3**

----- / -----

The setting out, and the answer are shown above. Simply start with the LHS and ask yourself, what is the first cubic number below the number 35. It is 27 or 3 cubed or 3^3.

So the first part of the 2 digit answer is a 3 which means the answer is "30 something".

"The something" is the second part of the answer on the RHS that determines the Unit's Digit. We look at the number "93**7**" but are only interested in the last digit which is the "7" shown in bold. We ask ourself, which cubic number, from 1 to 9, has its Unit's Digit or final digit ending in a "7". It can only be 27 or 3 cubed or 3^3.

Thus the final answer is 3/3 or 33.

EXERCISE 1

Find the Cubic Roots of the following perfect Cubic Numbers.
Do them mentally as One-Line Answers.

a) 185,193 =
b) 42,875 =
c) 1,331 =
d) 10,648 =
e) 157,464 =
f) 46,656 =
g) 970,299 =
h) 571,787 =
i) 110,592 =
j) 216,000 =

Answers to Exercise 1
a) 57
b) 35
c) 11
d) 22
e) 54
f) 36
g) 99
h) 83
i) 48
j) 60

How can you tell whether or not a number is a perfect cubic number.
There is a simple though unusual method, called "**Digit Sums**" which can solve this.
To determine a Digit Sum, add up all the individual digits of a number.
Complete the digit sum of the following numbers, (of Chart 1 shown before) in the spaces provided.
Eg: The Digit Sum of 64 is 6+4 = 10, but we have to keep going till we get to a single digit from 1 to 9, thus 10 = 1+0 = 1. Therefore, the Digit Sum of 64 = 1.

DIGIT SUM Of The CUBIC NUMBERS

Cubic Number	Summing of Digits	Final Digit Sum
1	1 =	1
8	8 =	8
27	2+7 =	9
64	6+4=10 =1+0 =	1
125		
216		
343		
512		
729		

You can therefore learn from this table, gazing at the RHS column, that The Sequence of Cubic Numbers has a Periodicity of 3, which means that the sequence of 3 numbers: 1,8,9 keep cycling or repeating.
That means if a number has its Digit Sum ending in a 1 or 8 or 9, then we know it is a perfect Cubic Number

eg 1: Is 456,533 a Perfect Cube?
Add its Digits and check. What is its cube root?

eg 2: Is 2,985,984 a Perfect Cube?
Add its Digits and check.

Answers: eg 1: 4+5+6+5+3+3= 8, therefore Yes. Its Cube Root is 77
eg 2: 2+9+8+5+9+8+4= 45= 4+5= 9, therefore Yes. Its Cube Root is 144

DAY 17

THE MAGIC OF 9

~ **Special Division By 9** eg: 1234 ÷ 9

~ Multiplication of Numbers by 9 (**Where the Multiplicand and the Multiplier have the same amount of Digits**)
eg: 345 x 999

~ Multiplication of Numbers by 9 (**Where the Multiplicand has a lesser amount of Digits than the Multiplier of Nines**)
eg: 34 x 999

~ **Decimalization of 1/19** (ie: 1 ÷ 19) Understanding Complements of 9 and utilizing the sutra: By One More. Remarkable 18 digit answer computed mentally!

THE MAGIC OF **NINE** IN **VEDIC** MATHEMATICS JAIN 2001

SPECIAL DIVISION BY NINE

Nine is indeed a special number. Look how easy it is to divide by 9.

EXAMPLE 1

> **34 ÷ 9**

eg: **34 ÷ 9** can be computed mentally in two steps.
1 – The first part of the answer is merely the first digit "3"
2 – The second part of the answer is the remainder which is the sum of the two digits "3+4" = 7.
Thus 34 ÷ 9 = 3 remainder 7 or abbreviated to 3 r 7.

What about division by nine with large numbers!

EXAMPLE 2

> **2401 ÷ 9**

Here is the solution illustrating the setting out:

```
9 )2 4 0  1
   2 6 6 r 7  where "r" is the remainder.
```

Step 1: The initial "2" is brought down unchanged:

```
9 )2 4 0 1
   ↓
   2
```

Step 2: The "2" is then added diagonally to the "4": which gives "6"

```
9 )2 4 0 1
   ↓↗
   2 6
```

Step 3: The "6" is then added diagonally to the "0": which gives another "6"

```
9 )2 4 0 1
   ↓↗↗
   2 6 6
```

Step 4: This "6" is then added diagonally to the "1":

```
9 )2 4 0 1
   ↓↗↗ ↗
   2 6 6  r7
```

Thus the Answer is 266 Remainder 7.

nb: The first part of the answer for 2401 ÷ 9 is the first number 2 and the remainder is the sum of the digits: 2+4+0+1 = 7

Try this Division, where there is a **Carry-Over involved**:

| EXAMPLE 3 |

| 2345 ÷ 9 |

Straight away, you know that the first part of the answer is the "2" and the remainder will be the sum of all the digits: 2+3+4+5 = 14. Really, this is 1 lot of 9 with a remainder of 5.

```
9 )2 3 4   5
   2 5 9 r 14
```

Since the remainder exceeds nine by 5, we need to add another "1" to the 259 and leave the remainder as 5.
= 260 r 5

| EXERCISE 1 |

Divide by 9:

a) 13

b) 33

d) 62

e) 64

continued...

f) 113

g) 314

h) 555

i) 666

j) 4102

k) 6321

l) 31021

m) 40004

The following example involves a few more steps:

EXAMPLE 4

$$888 \div 9$$

The working out will look like this:

```
9 )8 8  8
   9 8 r 6
```

If we put 8 for the first part of the answer, we will get 8+8=16 for the next step, so straight away we increase this first "8" to "9" which is the first part of the answer. We will add this "9" to the middle "8" to get "17". The "1" of the "17" has already been dealt with, so only put down the "7".
But we can forsee that this "7" will be added to another final "8" which involves another 2 figure or carry-over, so straight away increase this "7" to a "8".
"7+8" = 15 which gives a remainder of "1+5" = 6.
Thus the answer is 98 r 6.

EXERCISE 2

Divide by 9:

n) 3462

o) 8314

p) 5757

q) 334472

a) $13 \div 9 = 1 \, r \, 4$
b) $33 \div 9 = 3 \, r \, 6$
c) $50 \div 9 = 5 \, r \, 5$
d) $62 \div 9 = 6 \, r \, 8$
e) $64 \div 9 = 7 \, r \, 3$
f) $113 \div 9 = 12 \, r \, 5$
g) $314 \div 9 = 34 \, r \, 8$
h) $555 \div 9 = 61 \, r \, 6$
i) $666 \div 9 = 73 \, r \, 9 = 74 \, r \, 0 = 74$
j) $4102 \div 9 = 455 \, r \, 7$
k) $6321 \div 9 = 6/9/_11/r1+2 = 69+1/1+1/r3 = 70/2/r3 = 702 \, r \, 3$
l) $31021 \div 9 = 3,446 \, r \, 7$
m) $40004 \div 9 = 4,444 \, r \, 8$
n) $3462 \div 9 = 3/7/_13/r4+2 = 37+1/1+3/r6 = 38/4/r6 = 384 \, r \, 6$
o) $8314 \div 9 = 8/_11/2+1/r3+4 = 9/2/3/r7 = 923 \, r \, 7$
p) $5757 \div 9 = 5/_12/3+5/r8+7 = 5+1/1+2/8/r15 = 6/3/8/r15 = 639 \, r \, 6$
q) $334472 \div 9 = 3/6/_10/1+4/5+7/r3+2 = 3/7/1/5/_12/r3+2 = 3/7/1/6/3/r5$
$= 37,163 \, r \, 5$

◆◆◆

(The bible on Vedic Mathematics that launched a thousand websites.
The master and Priest Bharati Krsna Tirthaji, died in 1960, but this
book was published posthumously in 1967.
It was from this book that Kenneth Williams of the UK released the
first workbooks on Rapid Mental Calculation, revived this lost
subject, and then I released the first dvd in the world inspired by
this same book. "VEDIC MATHEMATICS FOR THE NEW MILLENNIUM"
See the link:
http://www.jainmathemagics.com/subcategory/19/default.asp).

MULTIPLICATION OF NUMBERS BY 9s
(Where the Multiplicand and the Multiplier have the same amount of digits)

Let the Multiplicand, the number being multiplied, be any number, and the multipliers will consist of 9s.
We will first consider examples where the Multiplicand and the Multiplier consist of the same number of digits, as in 345 x 999

EXAMPLE 5

345 x 999

The solution is formed from 2 parts:
1 – We subtract "1" from the 345, which gives the first half of the answer,
(In effect, we could call this "Lessening By One" a new sutra by the name of:
"By One Less" and
2 – We apply the sutra: All From 9 And The Last From 10 to the Multiplicand of 345.

This gives an instant answer of
= 345 – 1 / (9 – 3 / 9 – 4 / 10 – 5)
= 344 / 655
= 344,655

EXAMPLE 6

58,479 x 99,999

Apply "By One Less" to the Multiplicand of 58,479, giving the first half of the answer then tag on the sutra:
"All From 9 And The Last From 10 to the Multiplicand of 58,479."

This gives:
= 58479 – 1 / (9-5 / 9-8 / 9-4 / 9-7 / 10-9)
= 58478 / 41521
= 5,847,841,521

EXERCISE 3

Multiply the following numbers where the Multiplicand and the Multiplier have the same amount of digits:

a) 38 x 99

b) 67 x 99

c) 247 x 999

d) 831 x 999

e) 1,234 x 9,999

f) 5,678 x 9,999

g) 26,763 x 99,999

i) 357,089 x 999,999

j) 785,906 x 999,999

a) **38 x 99** = 38–1 / 9-3 / 10–8 = 37/6/2 = 3,762
b) **67 x 99** = 67–1 / 9-6 / 10–7 = 66/3/3 = 6,633
c) **247 x 999** = 247–1 / 9-2 / 9-4 / 10–7 = 246/7/5/3 = 246,753
d) **831 x 999** = 831–1 / 9-8 / 9-3 / 10–1 = 830/1/6/9 = 830,169
e) **1,234 x 9,999** = 1234–1 / 9-1 / 9-2 / 9-3 / 10–4 = 1233/8/7/6/6 = 12,338,766
f) **5,678 x 9,999** = 5678–1 / 9-5 / 9-6 / 9-7 / 10–8 = 5677/4/3/2/2 = 56,774,322
g) **26,763 x 99,999** = 26763–1 / 9-2 / 9-6 / 9-7 / 9-6 / 10–3 = 26762/7/3/2/3/7
   = 2,676,273,237
h) **94,726 x 99,999** = 94726–1 / 9-9 / 9-4 / 9-7 / 9-2 / 10–6 = 94725/0/5/2/7/4
   = 9,472,505,274
i) **357,089 x 999,999** = 357089–1 / 9-3 / 9-5 / 9-7 / 9-0 / 9-8 / 10–9
   = 357088/6/4/2/9/1/1 = 357,088,642,911
j) **785,906 x 999,999** = 785906–1 / 9-7 / 9-8 / 9-5 / 9-9 / 9-0 / 10–6
   = 785905/2/1/4/0/9/4 = 785,905,214,094

MULTIPLICATION OF NUMBERS BY 9s
(Where the Multiplicand has a lesser amount of digits than the Multiplier of 9s)

EXAMPLE 7

34 x 999

In this case, we need to append or add on another zero to the forepart of the "34" such that the working or setting out appears like this:
034 x 999
Now it is in the appropriate style as with the previous examples:

= 034 – 1 / (9-0 / 9-3 / 10-4)
= 033 / 966
= 33,966

EXERCISE 4

Multiply the following numbers where the where the Multiplicand has a lesser amount of digits than the Multiplier of 9s:

a) **8 x 99 =**

b) **6 x 99 =**

c) **24 x 999 =**

d) **83 x 999 =**

e) **234 x 9,999 =**

f) **678 x 9,999 =**

g) **267 x 99,999 =**

h) **94 x 99,999 =**

i) **357 x 999,999 =**

j) **85 x 999,999 =**

a) **8 x 99** = 08-1 / 9-0 / 10-8 = 7/9/2 = 792
b) **6 x 99** = 06-1 / 9-0 / 10-6 = 5/9/4 = 594
c) **24 x 999** = 024-1 / 9-0 / 9-2 / 10-4 = 23/9/7/6 = 23,976
d) **83 x 999** = 083-1 / 9-0 / 9-8 / 10-3 = 82/9/1/7 = 82,917
e) **234 x 9,999** = 0234-1 / 9-0 / 9-2 / 9-3 / 10-4 = 233/9/7/6/6 = 2,339,766
f) **678 x 9,999** = 0678-1 / 9-0 / 9-6 / 9-7 / 10-8 = 677/9/3/2/2 = 6,779,322
g) **267 x 99,999** = 00267-1 / 9-0 / 9-0 / 9-2 / 9-6 / 10-7 = 266/9/9/7/3/3
  = 26,699,733
h) **94 x 99,999** = 00094-1 / 9-0 / 9-0 / 9-0 / 9-9 / 10-4 = 93/9/9/9/0/6
  = 9,399,906
i) **357 x 999,999** = 000357-1 / 9-0 / 9-0 / 9-0 / 9-3 / 9-5 / 10-7
  = 356/9/9/9/6/4/3 = 356,999,643
j) **85 x 999,999** = 000085-1 / 9-0 / 9-0 / 9-0 / 9-0 / 9-8 / 10-5
  = 84/9/9/9/9/1/5 = 84,999,915

*"YOU NEVER CHANGE THINGS
BY FIGHTING THE EXISTING REALITY.
TO CHANGE SOMETHING,
BUILD A NEW MODEL
THAT MAKES THE EXISTING MODEL OBSOLETE."*

- Buckminster Fuller

(Art by Jain 108)

DECIMALIZATION OF 1/19

EXAMPLE 8

1 ÷ 19

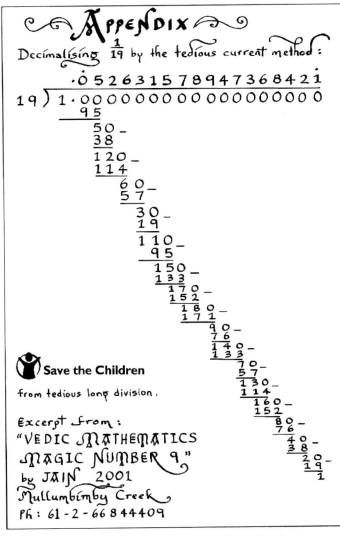

Demonstrate how this division (1 ÷ 19), that normally requires 35 lines of exhausting working out, can be done mentally, in your head, by understanding the Complementary Patternings of the Number 9. This particular example was Bharati Krshna Tirthaji's favourite example and it was used in the introduction of his of now world famous book: "Vedic Mathematics".

(1/19 is close to 1/20 which = .05, so we would expect our answer to be a bit more than .05).

Subtract the top Numerator of (1) from the bottom Denominator of (19). This gives us immediately how many digits in the required answer, which is 19 – 1 = 18 digits long.

= . _ _ _ _ _ _ _ _ _ _ _ _ _ _ _ _ _ _

We were never taught this in western schools. It is only particular to this case since 19 is a prime number, and this tells us that the last part of the answer is the numerator of 1.

We will now multiply this numerator of 1, beginning from right to left, by applying another sutra called: By One More Than The Previous Digit, which focusses on the denominator of 19, highlights the 1 of the 19 and multiplies this number (1) by the next counting number which is 2.

= . __ __ __ __ __ __ __ __ __ __ __ __ __ __ __ 1

So what we are about to perform is nothing more than multiplying 1 by 2 to arrive at high level mathematics that is normally achieved by a computer. Even the electronic calculator can only give 9 of the 18 digits in the final answer, so this method being explained now is far more efficient and sophisticated than calculators. We are bio-super-computers.

Remember the Doubling Sequence: 1 – 2 – 4 – 8 – 16 – 32 – 64 etc. This knowledge is now required.
But the setting out for this division, begins from the RHS.
All we are doing is simply doubling the last number, starting from the number 1, and setting out from the far right hand side of our page.
Notice as we multiply by 2, there are no carry-overs until we arrive to the number 8 giving 16 at which point we put down the 6 and carry over to the left hand side, the 1. Say, 1x2=2. Then 2x2=4. Then 2x4=8. We keep doubling the last part of the answer. It starts to look like this:

= . __ __ __ __ __ __ __ __ / __ __ __ __ $_1$6 8 4 2 1

Notice the introduction of the forward slash symbol " / " which acts as a place marker for the half waypoint of the calculation.

Then we say 2x6=12 plus the carry-over of 1 = 13, put down the 3 and carry-over the 1 to the left. We keep on continuing this process, until a magic gate opens, which is the original step of (19–1) giving 18, so when the number 18 arrives, we put down our pen, and complete the answer in our head, by using Pairs of Nine.
The mental calculations begin at this point, giving half of the 18 digit answer:

. __ __ __ __ __ __ __ __ / 9 4 7 3 6 8 4 2 1

Beginning from right to left, we subtract these 9 digits from 9, saying 9 minus 1 = 8, 9 minus 2 = 7, 9 minus 4 = 5 etc, literally pouring out the answer as fast as we can write:

$$= .0\ 5\ 2\ 6\ 3\ 1\ 5\ 7\ 8\ /\ 9\ 4\ 7\ 3\ 6\ 8\ 4\ 2\ 1$$

To complete the answer, one technical step is required, in the sense that this necklace having 18 beads, actually repeats this periodicity of 18 recurring digits infinitely, so the mathematician must place two dots, one above the "zero" on the far left hand side, and one above the "1" on the far right hand side, indicating clearly that the decimalization of the prime number 19 answer has a circulating phenomenon.

$$\frac{1}{19} = .\overset{\bullet}{0}526315789473684 2\overset{\bullet}{1}$$

(Of interest, this technique is also ambidextrous, meaning we can get this same answer had we divided the 1 by the 2 and begin to establish the left hand side of the answer: = .05 etc.
Quite Remarkable. This is the speed mathematics for the next generation of children).

EXERCISE 5

In the space below, repeat the following setting of instructions, to compute 1 ÷ 19 as a One-Line Answer.
This set of instructions is known as an **Algorithm**, or any step-by-step set of rules is an Algorithm, very similar to a cake recipe.

(Research: Find the name of the famous Arabic mathematician from whom this word "Algorithm" is derived from).

1 ÷ 19 =

. _ _ _ _ _ _ _ _ / _ _ _ _ _ _ _ _ _

DAY 18

Universal Sutra: "Vertically and Crosswise"

- Learning the Visual Pattern for All Multiplications
- **3 D**igits **x 2 D**igits: eg: 144 x 35 using Moving Multiplier Method
- Alternative Trachtenberg Method for 2 Digits x 2 Digits (2d x 2d)
- **4 D**igits **x 4 D**igits: eg: 1234 x 4321
- Worksheet on penciling in the actual "**X**" **shapes** up to
  5 **D** x 5 **D**igits

LEARNING MATHEMATICS BY PICTURES!
RIGHT-BRAIN, VISUAL or FEMININE BRAIN MATHEMATICS!

Now that you have learnt to multiply all your 2 Digits by Digits, by understanding a Picture (The "X-Shape), can you therefore predict what are the following "Shapes" for determining multiplication for 3 Digits By 3 Digits, 4 D x 4 D and 5 D x 5 D?

Fill in, by joining the dots, the appearance of the pattern used to solve Vertically and Crosswise, according to the number of digits being multiplied.
Be aware that the correct graphic solutions will involve a sense of symmetry or mirror axis around the center, akin to origami, in the sense that one side of the solution can be folded exactly over the other half.
The solution is shown on a previous worksheet.

It is quite spectacular that this diagram proves that **all Mathematical Multiplications can be solved by a picture**. It is the aim of this book to educate the world that this is the ultimate and most economical way to perform such multiplications.

Sutra: VERTICALLY AND CROSSWISE

EXERCISE 1

1 Digit x 1 Digit: o

                          o

2 Digits x 2 Digits:

o o   o o   o o

o o   o o   o o

3 Digits x 3 Digits:

o o o   o o o   o o o   o o o   o o o

o o o   o o o   o o o   o o o   o o o

4 Digits x 4 Digits:

o o o o   o o o o   o o o o   o o o o   o o o o   o o o o   o o o o

o o o o   o o o o   o o o o   o o o o   o o o o   o o o o   o o o o

5 Digits x 5 Digits:

o o o o o   o o o o o   o o o o o   o o o o o   o o o o o   o o o o o   o o o o o   o o o o o   o o o o o

o o o o o   o o o o o   o o o o o   o o o o o   o o o o o   o o o o o   o o o o o   o o o o o   o o o o o

TRACHTENBERG'S ALTERNATIVE
a MORE PREFERABLE **METHOD OF MULTIPLYING 23 X 81**

Multiplication of OUTER DIGITS + Multiplication of INNER DIGITS

EXAMPLE 1

23 x 81

Since Vedic Mathematics is considered a Mental, One-Line Arithmetic process, you may question the above method since it is not in one line. Why does the "81" get written underneath the "23"? Is it not more appropriate to write the question in one line: 23 x 81?
In fact, Trachtenberg, in his famous book: "The Trachtenberg Speed System of Basic Mathematics (1960) demonstrates such a one-line answer by correctly ascertaining the middle section (2x1)+(3x8) as multiplying the **Extreme Outer Digits** (2x1) [of **23 x 81**] and adding the multiplication of the **Inner Digits** (3x8) [of 2**3** x **8**1]. The units digit for the right side is obtained by multiplying the two right-side digits of 2**3** x 8**1** and the hundreds digit of the left-side is obtained by multiplying the two left-side digits of **2**3 x **8**1.

= 2x8 / ((2x1) + (3x8) / 3x1
= 16 / 26 / 3 which can be better written as: 16 / $_2$6 / 3
= 18/6/3 or **1,863**

In terms of mental swiftness and agility, I prefer to use this method as Vedic Mathematics tends to adopt the natural horizontal neural pathway. It's a simple and well packaged universal formula, this easy understanding of the Inner Digits and the Outer Digits. I support this method to be used as all other methods of Vertically and Crosswise train the student to set up in their Inner Mental Screen two lines of data:

23 x
81

whereas this book is a statement to start teaching the one-line approach: 23 x 81 as the superior mental system.

Ultimately, it is up to the student to adopt their preferred method.

Multiply the following 2 Digit by 2 Digit numbers, by either using the sutra: Vertically and Crosswise or by using the above Alternative Trachtenberg Method:

a) 14 x 17 =

b) 21 x 34 =

d) 42 x 65 =

e) 58 x 72 =

f) 64 x 91 =

g) 76 x 84 =

h) 85 x 43 =

a) **14 x 17 =** $1x1/(1x7)+(4x1)/4x7 = 1/11/28 = 1/13/8 = 2/3/8 = 238$

b) **21 x 34** $= 2x3/(2x4)+(1x3)/1x4 = 6/11/4 = 7/1/4 = 714$

c) **33 x 66** $= 3x6/(3x6)+(3x6)/3x6 = 18/36/18 = 18/37/8 = 21/7/8 = 2,178$

d) **42 x 65** $= 4x6/(4x5)+(2x6)/2x5 = 24/32/10 = 24/33/0 = 27/3/0 = 2,730$

e) **58 x 72** $= 5x7/(5x2)+(8x7)/8x2 = 35/66/16 = 35/67/6 = 41/7/6 = 4,176$

f) **64 x 91** $= 6x9/(6x1)+(4x9)/4x1 = 54/42/4 = 58/2/4 = 5,824$

g) **76 x 84** $= 7x8/(7x4)+(6x8)/6x4 = 56/76/24 = 56/78/4 = 63/8/4 = 6,384$

h) **85 x 43** $= 8x4/(8x3)+(5x4)/5x3 = 32/44/15 = 32/45/5 = 36/5/5 = 3,655$

ANSWER TO EXERCISE 1

MOVING MULTIPLIER METHOD For
3D X 2D and
4D x 2D etc

This unique method involves sliding the Multiplier starting from the RHS and moving along the left side. It uses the usual Cross-Multiplication of multiplying the diagonally aligned digits and adding those two sums.

EXAMPLE 2

> **1234 x 21**

You can see the 3 Cross-Multiplication steps involved:

```
1 2 3 4     1 2 3 4     1 2 3 4
    2 1       2 1       2 1
```

There are actually 5 steps involved when you include the Vertical Multiplication of the 2 extreme right digits and 2 extreme left digits. The 5 mental steps, beginning from Right to Left, are:

```
1 2 3 4 x
    2 1
--------
```

1×2 / $(1 \times 1)+(2 \times 2)$ / $(2 \times 1)+(3 \times 2)$ / $(3 \times 1)+(4 \times 2)$ / 4×1

$= 2 / 5 / 8 / \mathbf{1}1 / 4$

nb: when a double digit appears, as in the "11", it involves a carry-over, so it can be written as a "baby 1" which shown below as a subscript:

$= 2 / 5 / 8 / {}_11 / 4$
$= 2 / 5 / 9 / 1 / 4$
$= 25,914$

Multiply the following arrangements of 3, 4 or 5 digit numbers by these 2 digit numbers, using the Moving Multiplier Method:

a) 123 x
   21

b) 321 x
   32

c) 1221 x
   41

d) 1313 x
   22

e) 12234 x
   22

f) 32156 x
   11

a) 123 x 21 = 1x2/(1x1)+(2x2)/(2x1)+(3x2)/1x3 = 2/5/8/3 =**2,583**

b) 321 x 32 = 3x3/(3x2)+(2x3)/(2x2)+(1x3)/1x2 = 9/12/7/2 = 10/2/7/2
= **10,272**

c) 1221 x 41 = 1x4/(1x1)+(2x4)/(2x1)+(2x4)/(2x1)+(1x4)/1x1 =4/9/10/6/1
= 5/0/0/6/1 = **50,061**

d) 1313 x 22 = 1x2/(1x2)+(3x2)/(3x2)+(1x2)/(1x2)+(3x2)/3x2 = 2/8/8/8/6
= **28,886**

e) 12234 x 22 = 1x2/(1x2)+(2x2)/(2x2)+(2x2)/(2x2)+(3x2)/(3x2)+(4x2)/4x2
= 2/6/8/10/14/8 = 2/6/9/1/4/8 = **269,148**

f) 32156 x 11 = 3x1/(3x1)+(2x1)/(2x1)+(1x1)/(1x1)+(5x1)/(5x1)+(6x1)/6x1
= 3/5/3/6/11/6 = 3/5/3/7/1/6 = **353,716**

MULTIPLYING 4-DIGIT BY 4-DIGIT NUMBERS

Observe how we can extend the Pattern for

MULTIPLYING 3 -FIGURE NUMBERS

$$\vdots \quad \times \quad * \quad \times \quad \vdots$$

and progress to multiplying

4-FIGURE NUMBERS

$$\vdots \quad \times \quad * \quad * \quad * \quad \times \quad \vdots$$

| (7) | (6) | (5) | (4) | (3) | (2) | (1) |

like :

$$\begin{array}{cccc} 1 & 2 & 3 & 4 \\ 4 & 3 & 2 & 1 \end{array} \times$$

$$= 5,332,114$$

Mentally, some people like to work from the Left to the Right, but we will commence from the Right to the Left.
There are 7 steps required, shown as follows :

1) $(1 \times 4) = 4$

2) $(1 \times 3) + (2 \times 4) = 11$

3) $(1 \times 2) + (3 \times 4) + (2 \times 3) = 20$

4) $(1 \times 1) + (4 \times 4) + (2 \times 2) + (3 \times 3) = 30$

5) $(2 \times 1) + (4 \times 3) + (3 \times 2) = 20$

6) $(3 \times 1) + (4 \times 2) = 11$

7) $(4 \times 1) = 4$

This can be written linearly as :

$$= 4 \quad 11 \quad 20 \quad 30 \quad 20 \quad 11 \quad 04$$

$$= 5,332,114$$

Easy

4-FIGURE NUMBERS

step 7   step 6   step 5   step 4   step 3   step 2   step 1

Notice the Center Point at Step 4 which is a small "x" factor within the larger "X" factor. This means that of the 8 digits in the question, there will be 4 distinct Pairs or Cross-Multiplications to be considered.

NETWORK for MULTIPLYING 5-DIGIT by 5-DIGIT NUMBERS

Here is a network of lines that shows the creation of the pattern necessary to solve 5 Digits x 5 Digits, using the sutra: Vertically and Crosswise:

Following is the network for the multiplication of a five-digit number with another five-digit number.

Steps	Fixing 1st dot	Fixing 2nd dot	Fixing 3rd dot	Fixing 4th dot	Fixing 5th dot	Merging
1.						
2.						
3.						
4.						
5.						
6.						
7.						
8.						
9.						

(reproduced from "Calculations @ The Speed of Light" by Kranthi Kiran Tumma, 2002).

DAY 19
PYTHAGORAS' THEOREM

~ Definition
~ $3^2 + 4^2 = 5^2$
~ The 108 Triangle
~ Pythagorean Triples & Patterns
~ The Rope-Stretchers Knot of 12
~ Gnomon
~ The 3-4-5 Triangle In the Geometry of Flowers
~ Plato's "Most Beautiful Triangle"
~ 3-4-5 Proof by Mere Observation from Greece
~ 3-4-5 Algebraic Proof by Bhaskara, India & Proof By Tiling
~ Some Ancient Historical References on the 3-4-5 Triangle
~ Pythagorean derivation of the Musical Diatessaron or 3:4 Ratio
~ Derivation of the Symbols of the Master Mason
~ Using Pythagoras' Theorem to determine $\sqrt{2}$, $\sqrt{3}$ and $\sqrt{5}$ in the Vesica Piscis
~ Puzzle using Pythagoras' Theorem to determine the Shortest Path on The Cube

(Art by Jain 108)

"Geometry has two great treasures:
one is the Theorem of Pythagoras;
the other, the division of a line into extreme and mean ratio.
The first we may compare to a measure of gold;
the second we may name a precious jewel."
--Johannes Kepler

Definition:

The Pythagorean theorem: The sum of the areas of the two squares on the legs (*a* and *b*) equals the area of the square on the hypotenuse (*c*).

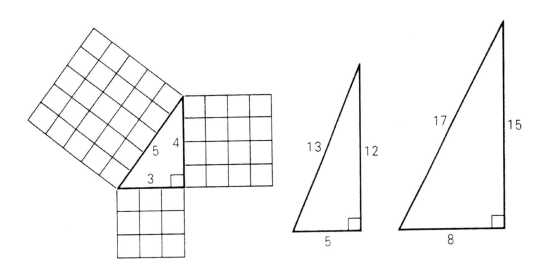

$$a^2 + b^2 = c^2$$
$$3^2 + 4^2 = 5^2$$

QUESTION:
If the base of a right-angled triangle is 5cm and the adjacent side is 12cm, what is the length of the longer side or hypotenuse?

SOLUTION:
Let "a" = 5cm, "b" = 12cm and the Unknown side, be called "c"
Therefore $5^2 + 12^2 = "c^2"$
Rearranging:
"c^2" = 25 + 144
"c^2" = 169
"c" = The square root of 169, which means, what number multiplied by itself gives 169?
Since 13 x 13 = 169, then the hypotenuse "c" = 13 square centimeters which is written as: $13cm^2$.

QUESTION: The 108 TRIANGLE.

If 3-4-5 is a Pythagorean Triple, it can be doubled to 6-8-10 and tripled to 9-12-15. Determine what are the other sides for the right angled triangle whose smallest side is 108 units.

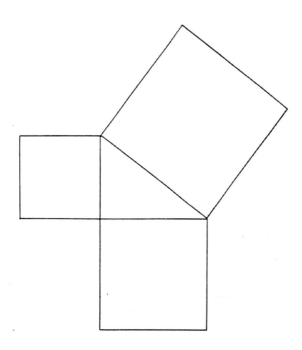

SOLUTION:

Let "a" = 108 units,   "b" = ?   and   "c" the hypotenuse = ?

Based upon the information of the standard 3-4-5, the smallest side of 3 units must be multiplied by 36 to reach a value of 108 units. This means that the other two sides of the 3-4-5 Triangle, "b" and "c", must be multiplied by the same factor of 36.

Therefore, "b" = 4 x 36. This can be solved mentally by applying Trachtenberg's Method of "1 Digit x 2 Digit" giving:

4x3/4x6 = 12/₂4 = 14/4 = 144 units

and "c" = 5 x 36. This too can be solved mentally by applying Trachtenberg's Method of "1 Digit x 2 Digit" giving:

5x3/5x6 = 15/₃0 = 18/0 = 180 units

Thus this is another Pythagorean Triplet: 108-144-180.

We can again double these numbers and arrive at:

216-288-360 an important triangle as it contains the number of degrees in a circle.

PYTHAGOREAN TRIPLES
are right-angled triangles where two sides differ by 1 unit,
as in 5-12-13   &   25-40-41 etc

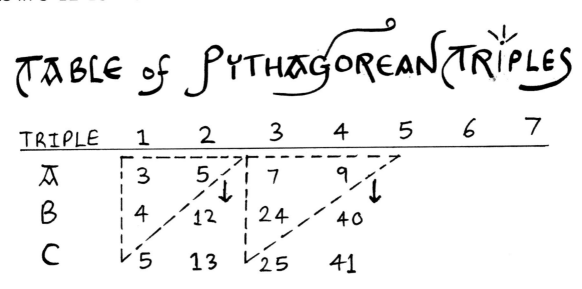

A Pythagorean Pattern Generating Pythagorean Triples

EXERCISE 1a

Check to see that (3, 4, 5) and (7, 12, 13) and (7, 24, 25) obey the
formula $a^2 + b^2 = c^2$

EXERCISE 1b

How would you create the 5th Triple?
This means you need to find the values of A, B and C.
First find A, in the Table.

How do you find the value of B?

If you know the value of B, then how do you get the value of C?

What are the 6th, 7th and 8th Pythagorean Triples?

ANSWERS 1a

(3, 4, 5)	$3^2 + 4^2 = 5^2$	$9 + 16 = 25$	Yes
(7, 12, 13)	$7^2 + 12^2 = 13^2$	$49 + 144 = 169$	Yes
(7, 24, 25)	$7^2 + 24^2 + 25$	$49 + 576 = 625$	Yes

ANSWERS 1b

"A" is the next consecutive odd number in the series,
ie: from 3-5-7-9 the next odd number is 11. ∴ A = 11
"B" is found by adding 3 numbers in an "L"-shape. Thus by adding
11+9+40 you get 60.
To get the "B" marked with an arrow, you need to add all 3 digits in
the triangle with dotted lines around it.
The value of "C" is always one more than the value of "B" thus
C=61.
Thus the 5th Triple is (11, 60, 61)

ANSWERS 1c

6th Triple has A = the next odd number which is 13
B = sum in the triangle of 11 + 13 + 60 = 84
C = one more than B which is 85
∴ The 6th Pythagorean Triple is **(13, 84, 85)**

7th **Triple** has A = the next odd number which is 15
B = sum in the triangle of 13 + 15 + 84 = 112
C = one more than B which is 113
∴ The 7th Pythagorean Triple is **(15, 112 ,113)**

8th **Triple** has A = the next odd number which is 17
B = sum in the triangle of 15 + 17 + 112 = 144
C = one more than B which is 145
∴ The 8th Pythagorean Triple is **(17, 144 ,145)**

Determine the missing number in this list of the first 12 primitive or distinct Pythagorean Triples: Let the missing number, the unknown, be written as "x".

a) (_, 4, 5)   =
b) (5, _, 13)   =
c) (7, 24, _)   =
d) (_, 15, 17)   =
e) (9, _, 41)   =
f) (11, 60, _)   =
g) (_, 35, 37)   =
h) (17, _, 145) =
i) (19, 180, _) =
j) (_, 84, 85)   =
k) (15, _, 113) =
l) (16, 63, _)   =

m) Verify the algebraic Proof of Pythagoras' Theorem, (known as Baudalayana's Theorem in the ancient Indian style), as demonstrated by the mathematician Bhaskara, (born in 1114AD) where the small central square has dimensions of (A-B).

Answers to Exercise 5

a) (3, 4, 5) since $x^2+4^2=5^2$, thus $x=\sqrt{(25-16)}=\sqrt{9}=3$
b) (5, 12, 13) since $5^2+x^2=13^2$, thus $x=\sqrt{(169-25)}=\sqrt{144}=12$
c) (7, 24, 25) since $7^2+24^2=x^2$, thus $x=\sqrt{(49+576)}=\sqrt{625}=25$
d) (8, 15, 17) since $x^2+15^2=17^2$, thus $x=\sqrt{(289-225)}=\sqrt{64}=8$
e) (9, 40, 41) since $9^2+x^2=41^2$, thus $x=\sqrt{(1681-81)}=\sqrt{1600}=40$
f) (11, 60, 61) since $11^2+60^2=x^2$, thus $x=\sqrt{(121+3600)}=\sqrt{3721}=61$
g) (12, 35, 37) since $x^2+35^2=37^2$, thus $x=\sqrt{(1369-1225)}=\sqrt{144}=12$
h) (17, 144, 145) since $17^2+x^2=145^2$, thus $x=\sqrt{(21,025-289)}=\sqrt{20,736}=144$
i) (19, 180, 181) since $19^2+180^2=x^2$, thus $x=\sqrt{(361+32,400)}=\sqrt{32,761}=181$
j) (13, 84, 85) since $x^2+84^2=85^2$, thus $x=\sqrt{(7,225-7,056)}=\sqrt{169}=13$
k) (15, 112, 113) since $15^2+x^2=113^2$, thus $x=\sqrt{(12,769-225)}=\sqrt{12,544}=112$
l) (16, 63, 65) since $16^2+63^2=x^2$, thus $x=\sqrt{(256+3,969)}=\sqrt{4,225}=65$

m) the solution is shown graphically 3 pages ahead.

QUESTION:

HOW MANY KNOTS NEEDED TO MAKE A RIGHT-ANGLE?

If you had a rope, how many knots would you need to make in the rope, to be able to use it, in a triangular form, for building purposes to determine a true right angle or corner of 90 degrees, as in a brick wall or wooden constructions.

(Answer on next page)

GNOMON:

Colour in the 3x3 squares below, and with the same colour, colour in the L-Shaped section to show the similarity of areas.
What does the word Gnomon mean?

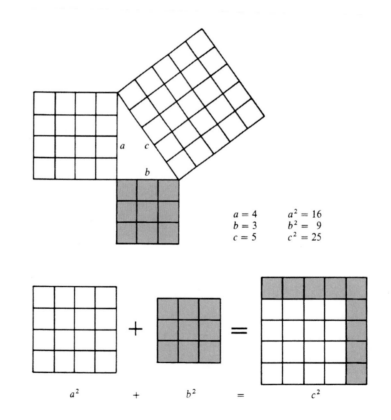

$$a = 4 \qquad a^2 = 16$$
$$b = 3 \qquad b^2 = 9$$
$$c = 5 \qquad c^2 = 25$$

$$a^2 \qquad + \qquad b^2 \qquad = \qquad c^2$$

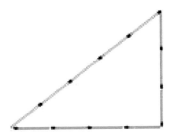

ANSWER:

HOW MANY KNOTS NEEDED TO MAKE A RIGHT-ANGLE?

Its called the **ROPE-STRETCHER'S KNOT** having 3+4+5 knots.
Mark these twelve knots upon the above Right-Angled Triangle.

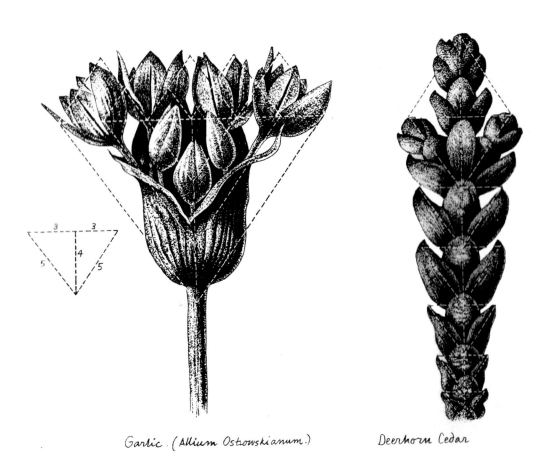

Garlic. (Allium Ostrowskianum.)   Deerhorn Cedar

• The 3-4-5 Triangle In the Geometry of Flowers, specifically:
Garlic on the left, and DeerHorn Cedar on the right.
(images taken from Gyorgy Doczi's classic: "The Power Of Limits").

3 PROOFS:

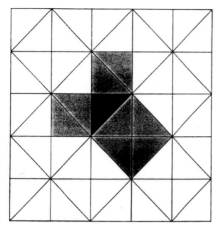

105. Ancient Greek proof of Pythagorean theorem
for the isosceles right triangle

1 • 3-4-5 Proof in the Greek style, Proof by Mere Observation!

Show the proof below of Pythagoras' Theorem in terms of A & B.

An elementary example might be the proof of Pythagoras' theorem given by the Indian mathematician Bhāskara (born A.D. 1114). He simply draws four equal right-angle triangles as in figure 6.4. The area of each triangle is $ab/2$, so c^2 (the area of the

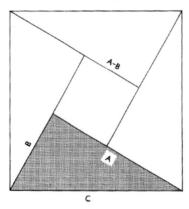

Fig. 6.4. Pythagoras' theorem

square) is equal to $4ab/2 + (a - b)^2 = a^2 + b^2$. The reader can easily verify the construction.

2 • 3-4-5 Algebraic Proof in the ancient Indian style, as demonstrated by the author Bhaskara, (born in 1114AD)

$$c^2 = 4 \times \tfrac{1}{2}ab + (a-b)^2$$
$$= a^2 + b^2$$

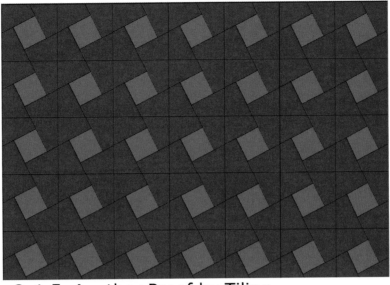

3   • 3-4-5 Another Proof by Tiling

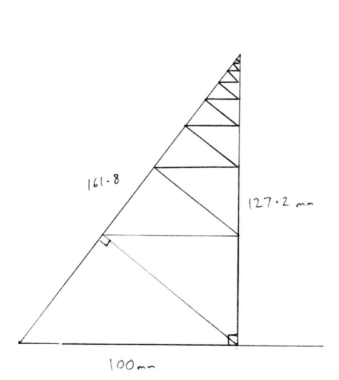

161·8

127·2 mm

100mm

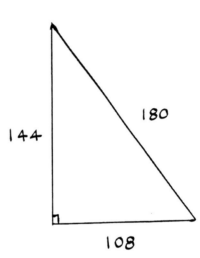

144

180

108

Higher Harmonic Octaves
of the 3-4-5 Right-
Angled Triangle:
$108^2 + 144^2 = 180^2$

• Plato's "**Most Beautiful Triangle**" showing "The Stairway To Heaven" or "Jacob's Ladder" based on the Infinite Series of the Divine Phi Proportion (1:1.618033…. and the Square Root of Phi 1.272). This triangle can be simplified to **1 : Square Root Of Phi : Phi**
or **1 : 1.272 : 1.618**

SOME ANCIENT HISTORICAL REFERENCES
ON THE 3-4-5 TRIANGLE
IN EGYPTIAN, CHINESE, JEWISH, BABYLONIAN, ARABIC, GREEK AND MASONIC CULTURES.

The "Forty-seventh Problem" was among the Ancient Egyptians the symbol of Osiris, Isis and Horus.

Clay Tessera found in an Egyptian Tomb.

- 3-4-5 Triangle (Euclid's 47ᵗʰ Proposition) shown in Egyptian lore.

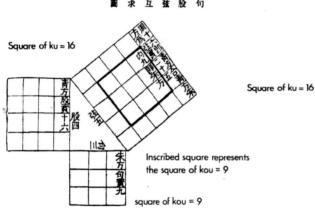

Square of ku = 16

Inscribed square represents the square of kou = 9

square of kou = 9

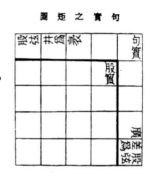

Square of ku = 16

Ancient Chinese mathematical problems dealing with the gnomonic principle.

71

- 3-4-5 Triangle in Chinese, showing the concept of Gnomon!

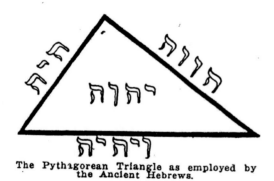

The Pythagorean Triangle as employed by the Ancient Hebrews.

- 3-4-5 Triangle in Hebrew, showing the sacred names of God, and the 3-4-5 Triangle in Babylon.

A rare appearance of the 3-4-5 Triangle in Arabic:

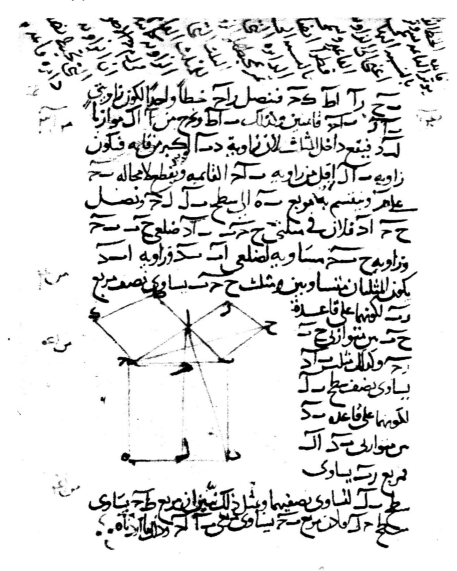

Page from an Arabic commentary on Euclid (c. A.D. 1250). After their assimilation by the Moslems, the thirteen books of Euclid's Elements exercised a more powerful influence on European mathematics than almost any other treatise of ancient Greece.

Pythagoras' Theorem, in Arabic, as recorded in Euclid's Element: (This diagram, and the next one, have been reproduced from Lance Hogben's book: "Mathematics In The Making").

Proposition 47. Theorem.

In a right-angled triangle the square described on the hypotenuse is equal to the sum of the squares described on the other two sides.

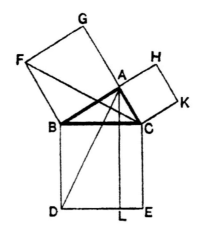

Let ABC be a right-angled triangle, having the angle BAC a right angle:
then shall the square described on the hypotenuse BC be equal to the sum of the squares described on BA, AC.

Construction. On BC describe the square BDEC; I. 46.
and on BA, AC describe the squares BAGF, ACKH.
Through A draw AL parallel to BD or CE; I. 31.
and join AD, FC.

Proof. Then because each of the angles BAC, BAG is a right angle,
therefore CA and AG are in the same straight line. I. 14.

Now the angle CBD is equal to the angle FBA,
for each of them is a right angle.
Add to each the angle ABC:
then the whole angle ABD is equal to the whole angle FBC.

Proposition 47,
originally coined by Euclid for Pythagoras' Theorem,
from his masterpiece "Elements"
which when translated into Greek, became the template
for all modern mathematical high school books in the world.

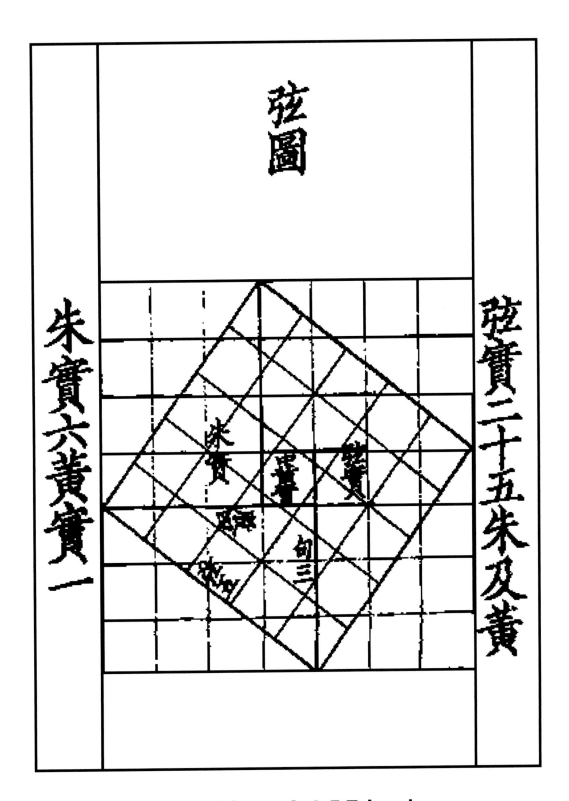

Hsuan-Thu Oldest Chinese 3-4-5 Triangle

3-4-5 Triangles embedded in a 7x7 matrix, cleverly shows the precise distances of the sides being 3 units, 4 units and 5 units.

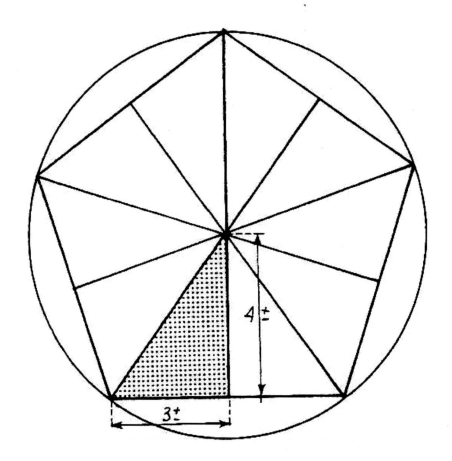

Diatessaron — or 3:4 – corresponds to the pentagon's triangle, approximated by a 3-4-5 triangle. 3:4 = 0.75

• 3-4-5 Triangle in Music and the Pentagon, showing the ancient Pythagorean derivation of the Diatessaron or the approximate 3:4 musical ratio.

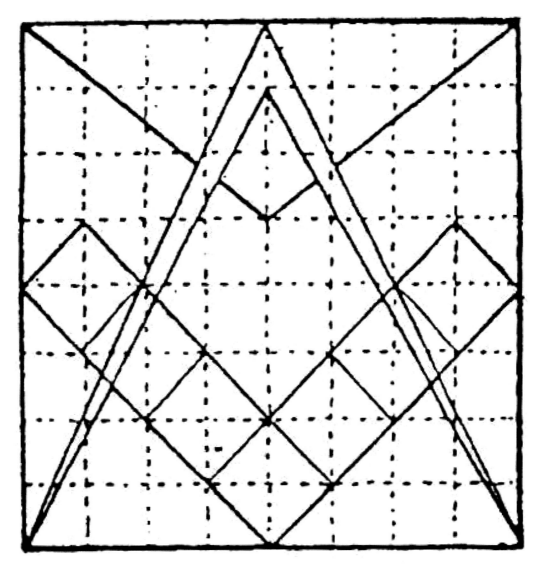

Derivation of Symbols of Master Mason

• 3-4-5 Triangle shown mystically in the 8x8 Grid or Chess-Board-like array displaying the Masonic symbols of the Set-Square and Compass.

The numbers 3 & 4 & 5 expressed geometrically as a Triangle, as a Square and as a Pentagon are all shown in the Vesica Piscis (union of 2 Circles passing thru their two centres).

The **Vesica Piscis** is **The Mother of All Form**, and exhibits the important Root Harmonics of Root 2, Root 3 & Root 5. Show how $\sqrt{2}$ (=1.4142… the diagonal of the Square) is derived; show how $\sqrt{3}$ (=1.732… the bisection of the Equilateral Triangle) is derived; show how $\sqrt{5}$ (=2.236… the diagonal of the Double Square and an integral part of the formula for the Divine Proportion $[1+\sqrt{5}] \div$ by 2 = 1:1.618033…) is derived. The diagrams below will assist you.

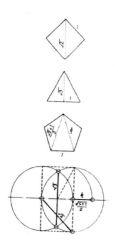

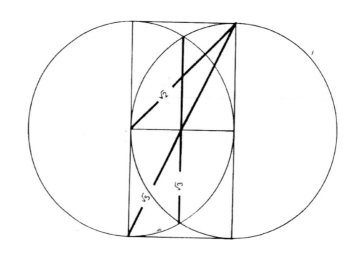

Calculating the length of the *vesica piscis*

Use the Pythagorean right-angled triangle COB and assume the radius = 1.
Hypotenuse $CB^2 = OB^2 + CO^2$
Therefore $OB^2 = CB^2 - CO^2$
$OB^2 = 1^2 - (\frac{1}{2})^2$
$OB^2 = 1 - \frac{1}{4} = 0.75$
$OB = \sqrt{0.75} = 0.8660254$
Therefore, $AB = 2 \times OB = 2 \times 0.8660254 = 1.7320508$
Which just happens to be $\sqrt{3}$.
So the length AB of the *vesica piscis* is $\sqrt{3}$.

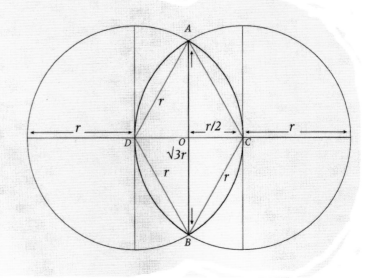

2 PUZZLES
1. What is the Shortest Path on the Cube ?
2. What is the Distance of the Space Diagonal ?

PUZZLE 1:

Can you work out the shortest distance or pathway between two points marked A & G on the Cube whose common edge length is 1m?

In 3-Dimensions, A & G are the end points of an opposing space diagonal.

The only condition regarding movement is that you can only travel over the outer flat surfaces, not interiorly through 3-Dimensional space.

Hint: you will have to test your choice of distances using Pythagoras's Theorem.

Keep a record of all distances measured for comparative reasons.

Let **O** be the **O**rigin or Centre of the Cube
Let **AG** = the space diagonal
AB = 1 unit
AG = AO + OG
M, N, P, Q and I, J, K, L are Midpoints of an Edge

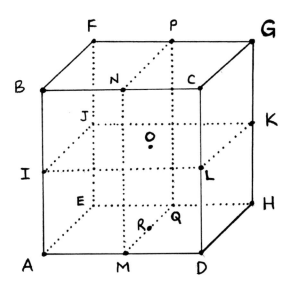

Here are some possibilities for your contemplation:

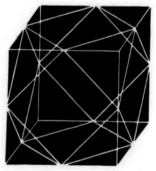

(Iona Miller's Design of some
possible Cubic Pathways)

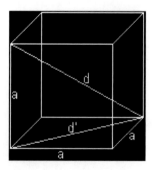

Use the space on this page to work out some of the distances:

PUZZLE 1: shortest distance from A to G, moving only upon the surface.

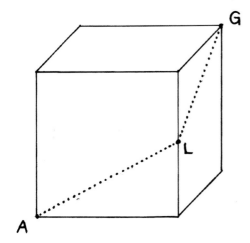

This is the visual solution, but first you would have had to know the distances of only AC and AL, then compare the different pathways. There are though 3 options:

Option 1: AD + DC + CG = 1 + 1 + 1 = 3 units which is the longest path.

Option 2: AC + CG = 1.414 + 1 = 2.414 units which is the second best option, and

Option 3: AL + LG = 1.118 + 1.118 = 2.236 = √5 which is the shortest and most efficient path.

nb: **√5** is part of the most famous formula for the Divine Proportion Phi (φ)= (1+√5) ÷ by 2 which = the Phi Ratio of 1:1.618033... Since Phi is really the aesthetics of Mathematical Beauty, it is appropriate that the shortest path on the exterior of the cube is solved by a distance of √5.

$$AC = \sqrt{[(1)^2 + [(1)^2]}$$
$$= \sqrt{2}$$
$$= 1.414...$$

$$AL = \sqrt{[(1)^2 + [(1/2)^2]}$$
$$= \sqrt{(5/4)}$$
$$= \sqrt{5}/2$$
$$= 2.236 \div by\ 2$$
$$= 1.118...$$

ANSWER PUZZLE 2
What is the Distance of the Space Diagonal ?

Let **O** be the **O**rigin or Centre of the Cube
Let **AG** = the space diagonal
Let **R** = the midpoint of the Square Base

How to measure AG?
AG Space Diagonal = 2 x AO
First plot a point R at the very base of the cube, in the centre of the Square Base ADEH.
R is the midpoint of MQ. This will help us measure AO and the Space Diagonal AG.
To know AO, we have to determine the length of AR first. Visualize the right-angled triangle RAM. Using Pythagoras Theorem,

$$AR = \text{Square Root of } [(1/2)^2 + [(1/2)^2]$$
$$= \sqrt{1/2}$$
$$= 1/\sqrt{2}$$

Now visualize spatially, the vertically standing right-angled triangle of ARO

$$AO = \sqrt{[(1/\sqrt{2})^2 + [(1/2)^2]}$$
$$= \sqrt{(1/2 + 1/4)}$$
$$= \sqrt{(3/4)}$$
$$= \frac{\sqrt{3}}{2}$$

∴ since AG = 2 x AO

then $AG = 2 \times \dfrac{\sqrt{3}}{2}$

$= \sqrt{3}$
$= \textbf{1.732...}$

Thus in any Unit Cube, the interior Space Diagonal is $\sqrt{3}$ and the surface Face Diagonal is $\sqrt{2}$

DAY 20

- Main and Final **Revision Test called "MATHLETICS"** with **Answers**
- Feedback Sheet
- **End of Course**
- Offering of **"Certificate Of Completion"**

- **APPENDIX**
  - ~ Some **Products by Jain**
  - ~ **Extra Puzzles** on **Combinatorics** (14 Examples)
  - ~ **1895 8ᵗʰ Grade Final Exam**

END Of RAPID MENTAL CALCULATION
LEVEL ONE
(JAIN MATHEMAGICS ONLINE CURRICULUM
For The GLOBAL SCHOOL)
For TEENAGERS From AGES 13 To 19

END OF COURSE

Issuing of laminated **CERTIFICATES** to STUDENTS.

Celebration

(Art By Jain, 1991)

MATHLETICS TEST: Level 2

RULES or GUIDELINES for the FINAL MATHLETICS TEST:

This final exam or test is not meant to be about competitiveness, it is offered as a mean to create classroom fun, and putting the students on focus in regard to remembering all their sutras and formulae and how to arrive with mental one-line answers or solve some geometrical problem.

• Before calling out the answer, first identify the name of the Sutra or Method of Solving, then give the answer.
• One point is given for the Naming of the Sutra,
• One point is given for the Correct Answer.
• Select a person in the room as Score-Keeper.
• In the classroom situation, once you have called out an answer, and it is not correct, you are not allowed to call out a second time; this gives the other students a chance to call out the answer. This also invites you to carefully think of your answer before calling it out.
• The questions will be selected in chronological order from the beginning of the book till the end, systematically going through Day 11 to Day 12 etc to Day 19.
• The prize for the student with the highest score will receive 2 dvds valued at $90 by Jain Mathemagics.... One called "Vedic Mathematics For The New Millennium", (the first complete dvd in the world on this topic) and an Introduction to Ancient Knowledge called "The Living Mathematics Of Nature", 1 of 5 of a 5 dvd set.

nb: This Test can be done as a written test so students are not interrupted, or as an Oral Test in Public with Parents viewing...

Revision Test: "MATHLETICS"

No.	NUMBER QUESTION	RELEVANT SUTRA	ANSWER
			DAY 11
1	Recite the Binary Sequence up to 1024 ?		
2	What is the 10thTriangular Number (TN10) ?		
3	Using Pascal's Triangle, what is 11^3 ?		
4	Write the Sequence of Fibonacci Numbers up to 233 ?		
			DAY 12
5	What is 30% of 370 ?		
6	What is 53% of 5300 ?		
7	Express the fraction 7/42 as a percentage to 2 Decimal Places ?		
			DAY 13
8	Decimalize the fraction 1/8 ?		
9	Decimalize the fraction 1/40 ?		
10	Decimalize the fraction 1/11 ?		
11	Decimalize the fraction 1/7 ?		
12	What value is half way between 2.6 and 2.65 units		

13	Express 7/12 as an Egyptian Fraction ?		
14	What is the formula for the Difference of Two Squares $(a^2 - b^2) =$		**DAY 14**
15	What is $14^2 - 12^2$?		
16	What is the square root of the sum of the first 9 odd numbers ?		
17	What is 325^2 ?		**DAY 15**
18	What is the square root of 4096 ?		
19	Using the Stroke Method of Multiplication, what is 12x24 ?		
20	What is 7x7x7 or 7^3 using Trachtenberg Method ?		**DAY 16**
21	What is $_3\sqrt{35,937}$ ie: the cube root of 35,937 ?		
22	What is $2401 \div 9$?		**DAY 17**
23	What is 345 x 999 ?		
24	Solve 23 x 81 using the Alternative Trachtenberg Method ?		**DAY 18**

25	Using the Moving Multiplier Method, what is 321 x 32 ?		
26	Determine the missing length in this Pythagorean Triple: (5, __, 13) ?		DAY 19
27	What is the length of the diagonal of the Double Unity Square ?		
28	Can you calculate or remember the length of the Vesica Piscis, based on 2 intersecting circles whose radii = 1 unit ?		
	Fini		

CONCLUSION:

Thus ends the second part of this course, Level Two, taking us closer to a deeper matheological understanding of how the Universe works. (Matheology = ma**THEO**logy. "Theo" is an ancient Greek word meaning "God".

Revision Test: "MATHLETICS Level 2"
ANSWERS

No.	NUMBER QUESTION	RELEVANT SUTRA	ANSWER
1	Recite the Binary Sequence up to 1024 ?		**DAY 11** 1 - 2 - 4 - 8 - 16 - 32 - 64 - 128 - 256 - 512 - 1024
2	What is the 10thTriangular Number (TN10) ?	"By One More"	$= (10 \times 11) \div 2$ $= 55$
3	Using Pascal's Triangle, what is 11^3 ?		$\begin{array}{c} 1 \\ 1 \ 2 \ 1 \end{array}$ $= 1 \ 3 \ 3 \ 1 \quad = 1{,}331$
4	Write the Sequence of Fibonacci Numbers up to 233 ?	Starting from 1,1 add the preceding number	1 - 1 - 2 - 3 - 5 - 8 - 13 - 21 - 34 - 55 - 89 - 144 - 233
5	What is 30% of 370 ?	$\dfrac{30}{100} \times 370$	**DAY 12** $= 111$
6	What is 53% of 5300 ?	Squaring of Numbers In The Fifties	$= 53^2 \times 10$ $= (25+3 \ /3^2) \times 10$ $= 2809 \times 10$ $= 28{,}090$
7	Express the fraction 7/42 as a percentage to 2 Dec Pl ?		$= 1/6 \times 100$ $= .1666 \times 100$ $= 16.66\%$

8	Decimalize the fraction 1/8 ?		**DAY 13** $= ½ \times ½ \times ½$ $= .5 \times .5 \times .5$ $= .125$
9	Decimalize the fraction 1/40 ?		$= 1/10 \times ¼$ $= .1 \times .25$ $= .025$
10	Decimalize the fraction 1/11 ?		$= .0909090909$ $= .09$ repeater
11	Decimalize the fraction 1/7 ?		$= .142857142857$ $= .142857$ repeater
12	What value is half way between 2.6 and 2.65 units ?		$= (2.60 + 2.65) \div 2$ $= 5.250 \div 2$ $= 2.625$
13	Express 7/12 as an Egyptian Fraction ?		$= 1/3 + ¼$
14	What is the formula for the Difference of Two Squares $(a^2 - b^2) =$		**DAY 14** $(a^2 - b^2) = (a-b)(a+b)$
15	What is $14^2 - 12^2$? Use the above formula.		$= (14 - 12) \times (14 + 12)$ $= 2 \times 26$ $= 52$

			cont...
16	What is the square root of the sum of the first 9 odd numbers ?		$= \sqrt{(1+3+5+7+9+11+13+15+17)}$ $= \sqrt{81}$ $= 9$
17	What is 325^2 ?		**DAY 15** $= 3^2 \,/\, 3{\times}5 \,/\, 25^2$ $= 9 \,/\, {}_15 \,/\, 625$ $= 10/5/625 \qquad = 105{,}625$
18	What is the square root of 4096 ?		$= 40/96 = 6 \,/\, \_ = 64$ or 66. Average $= 6{\times}7 = 42$ which is >40 $\therefore$ choose the lesser option $= 64$
19	Using the Stroke Method of Multi-plication, what is 12x24 ?		$= 2\,/\,8\,/\,8$ $= 288$
20	What is 7x7x7 or 7^3 using Trachtenberg Method ?		**DAY 16** $= 49{\times}7$ $= (7{\times}4)/(7{\times}9)$ $= 28/{}_63$ $= 34/3 \quad = 343$
21	What is $_3\sqrt{35{,}937}$ ie: the cube root of 35,937 ?	Vilokanam "By Mere Intuition"	$= 35 \,/\, 937$ $= \dfrac{3^3}{3} \,/\, \dfrac{(10 - 7)}{3} \qquad = 33$
22	What is $2401 \div 9$?	Special One-Line Division by 9	**DAY 17** $= 9\,)\underline{2\;4\;0\;\;1}$ $\quad\;\; 2\;6\;6\;r\,7$ $= 266$ remainder 7
23	What is 345 x 999 ?	"By One Less"	$= (345 - 1) \,/\, (9{-}3)(9{-}4)(10{-}5)$ $= 344 \,/\, 655 \qquad = 344{,}655$

24	Solve 23 x 81 using the Alternative Trachtenberg Method ?	"Inner Digits by the Inner Digits & Outer by Outer"	**DAY 18** = 2 x 8 /(2x1) + (3x8) / 3 x 1 = 16 / $_2$6 / 3 = 18 / 6 / 3 = 1,863
25	Using the Moving Multiplier Method, what is 321 x 32 ?	Vertically & Crosswise	= 10,272
26	Determine the missing length in this Pythagorean Triple: (5, __, 13) ?	Pythagoras' Theorem	**DAY 19** = 12 = (5, 12, 13) since $5^2+x^2=13^2$, thus x=$\sqrt{(169-25)}$ =$\sqrt{144}$ =12
27	What is the length of the diagonal of the Double Unity Square ?	Pythagoras' Theorem	= $\sqrt{5}$ = since it is a 1x2 rectangle $x^2 = 1^2 + 2^2$ thus x=$\sqrt{(1 + 4)}$ =$\sqrt{5}$
28	Can you calculate or remember the length of the Vesica Piscis, based on 2 intersecting circles whose radii = 1 unit ?	Pythagoras' Theorem	= $\sqrt{3}$ = since there is a right-angled triangle $x^2 = 1^2 - (1/2)^2$ thus x=$\sqrt{(1 - 1/4)}$ =$\sqrt{3}/2$ but there are 2 of these vertical lengths required, thus x2 = $\sqrt{3}$
	Fini		

FEEDBACK SHEET

Name and Age:

....

What is your Profession?

....

Date:.......................................

Name of Seminar(s):   Vedic Mathematics aka Rapid mental Calculation
Magic Squares
Divine Phi Proportion
3-Dimensional Geometry

How would you rate this Seminar? (Please Circle: 1 is low, 10 is excellent)

1   2   3   4   5   6   7   8   9   10

What did you enjoy most about this Seminar?
Were there any special moments or peak experiences?

What criticism or improvement could you recommend to Jain to help in upgrading his next presentation?

Do you give permission that Jain can use any of the above short feedbacks as a TESTIMONIAL that can posted on Jain's website?

Yes   No

Is there another person who you would like Jain to contact to recommend this course to. I will need their:

Name:
Phone Number:
Nearest Main City of Residence:
Email:

END OF Level 2 COURSE

Issuing of laminated **CERTIFICATES** to STUDENTS

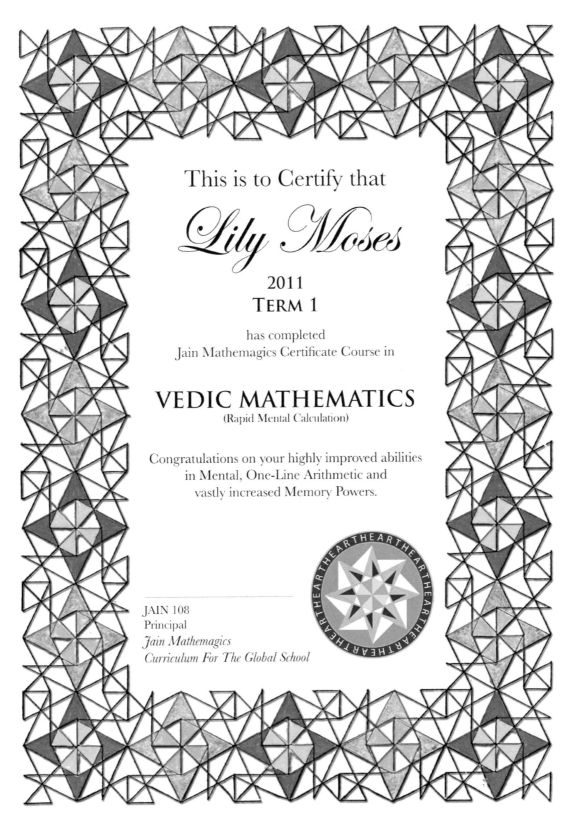

This is to Certify that

Lily Moses

2011
TERM 1

has completed
Jain Mathemagics Certificate Course in

VEDIC MATHEMATICS
(Rapid Mental Calculation)

Congratulations on your highly improved abilities
in Mental, One-Line Arithmetic and
vastly increased Memory Powers.

JAIN 108
Principal
Jain Mathemagics
Curriculum For The Global School

APPENDIX

JAIN MATHEMAGICS and VEDIC MATHMATICS
DUAL or COMBINED LEARNING CLASSES
FOR TEENS and PARENTS (ages 13 and above)
FOR JUNIORS and PARENTS (ages 9 to 12)
For the 4 School Terms of 2011

> ➢ Get involved with your Child's education
> ➢ Learn the supreme Art of Mental Calculation
> ➢ Develops your Mathematical Confidence
> ➢ Increases your Memory Power
> ➢ Dr Jain heals past Mathematical Traumas.

(for more information about the classes, click on:
http://www.jainmathemagics.com/page/3/default.asp)

The Living Mathematics Of Nature

Distributed by Jain Mathemagics

5 DISC DVD SERIES

New DVD Series

Introduction to Jain Mathemagics
An introduction to the 4 topics below.
Explains each topic, providing good overview of work.

Vedic Mathematics
Rapid Mental Calculation.
No more calculators. Increases your memory power!

Magic Squares
Translating numbers into Atomic Art.
Teaches children pattern recognition.

The Divine Phi Proportion
Explores the Geometry of Flowers
identical to the Human Canon!

3-Dimensional Geometry
The 5 Platonic Solids
adored by Pythagoras and his community.

Distributed by **JAIN MATHEMAGICS**
777 Left Bank Road Mullumbimby
NSW 2482 Australia

Ph: (02) 6684 4409
Email: jain@jainmathemagics.com
www.jainmathemagics.com

Videos & Books Available Online...

www.jainmathemagics.com

Copyright © Jain Mathemagics 2005

"omnia apud me mathematica fiunt:
With me everything turns into mathematics."
Descartes

Jain MatheMagics

VEDIC MATHEMATICS SACRED GEOMETRY
MAGIC SQUARES ANCIENT KNOWLEDGE

 **Jain's Interactive Workshops will help you discover**

Maths as Art

Maths as Science

Maths as History

Find out what you didn't learn in school...

Vedic Mathematics
(Rapid Mental Calculation)

Find out how to accomplish complex and rapid calculations...in your head!

Used by NASA but not released to the public, Vedic Mathematics is on the leading eday of rediscovered technology - an ancient yet simple system of mental one-line arithmetic.

Jain will show you the 16 basic Sutras - Word Formulae - which can solve all known mathematical problems.

You will learn to multiply 2-digit x 2 Digit numbers, like 24 x 24 in your head... with no calculator, pen or paper.

Magic Squares
(Translating Numbers into Atomic Art)

The Ancients discovered internal symmetries and harmonies that lay hidden within the chaos.. elegant and PHi-Ratioed patterns,.. atomic structures often symbolic of life's journey.

The Magic Square of 3x3, known as the Lo-Shu, in ancient China, is the centre of the Sino/Tibetan Cosmology. The sums of the columns, rows and diagonals all add up to 15, symbolically creating Order amidst the Chaos, and Equality in all dimensions.

Jain will show you how to create complex and intricate atomic art from numbers in various Magic Squares.

Three-Dimensional Sacred Geometry
(5 Platonic Solids)

Explore multi-dimensional geometry and the sub-atomic realm.. Jain will help you grasp the concepts of a Fourth Dimension and Time Travel.

Discover the importance of the Star Tetrahedron's 24 faces and 24 edges, and how this links to the 24 edges of the Cube Octahedron...the compressible and alchemical shape revered by Buckminster Fuller. Glimpse hidden mysteries within the 5 Platonic Solids and the 13 Archimedean Solids. Learn the 3-Dimensional form of the 5-pointed star that Pythagoras called the fifth element - Spirit aka Dodecahedron.

Learn the true value of Pi, based on Fractal Compression: Jain Pi = 3.14460..

Divine Proportion
(Lost Secrets of the 108 Phi Code)

Explore ancient hidden mysteries with Jain... why the shape of the pine cone is optically similar to the human heart, why we are awestruck when we view Sacred Architecture like the Parthenon of Greece and the Pyramids of Giza, and why we are attracted to the famous Mona Lisa.

Discover the secret codes concealed in the Gayatri Mantra, the most famous Eastern prayer for Enlightenment

The human body is in resonance with the Living Mathematics of Nature. Embrace part of the Ascension process already morphologically and geometrically encoded into our bio-magnetic and bio-electrical fields.

For more information, contact:
jain@jainmathemagics.com   www.jainmathemagics.com

Jain 108 Mathemagics
Sacred Geometry Mystery School

- Level 1 . Beginners Course   5 Days of Remembering
- Level 2 . Advanced Course   5 Days of Awakening
- Level 3 . Teacher Training   5 Days of Mastery

Art of Number

Level 1 : • Digital Compression of the Multiplication Table
revealing Atomic Structure of Rutile + Platinum Crystal

Level 2 : • Prime Numbers 24 Pattern + 4th Dim. Templar Cross
• Binary Code   origin of VW Symbol

Rapid Mental Calculation

Level 1:
• Multiplicat...
• Magic Fingers
Level 2:
• One Line Division
• Square + Cube Roots
• Fractions

Magic Squares

Level 1:
• Magic Sqs. of 3, 4, 5, 6, 7
Level 2:
• Magic Squares of 8, 9, 10, 11, 12, 16
• Magic Cubes + Stars

Divine Phi Proportion (108 Codes)

Level 1: • Phi Spiral
• Fibonacci Numbers
• 108 Phi Code 1
Level 2: • Pentagram Constr.
• 108 Phi Code 2
• Mystical Squaring of Circle
• Vesica Pisces
(the Mother of all Form)

3-Dimensional Sacred Geometry

Level 1:
• the 5 Platonic Solids
• Fold Up Net of Dodecahedron
Level 2:
• the 13 Archimedean Solids
• Cuboctahedron perfected
• Truncated Icosahedron (Soccer Ball Shape).

JAIN F.R.E.E.D.O.M.S Non Profit Org...
trading as: JAIN MATHEMAGICS
www.jainmathemagics.com

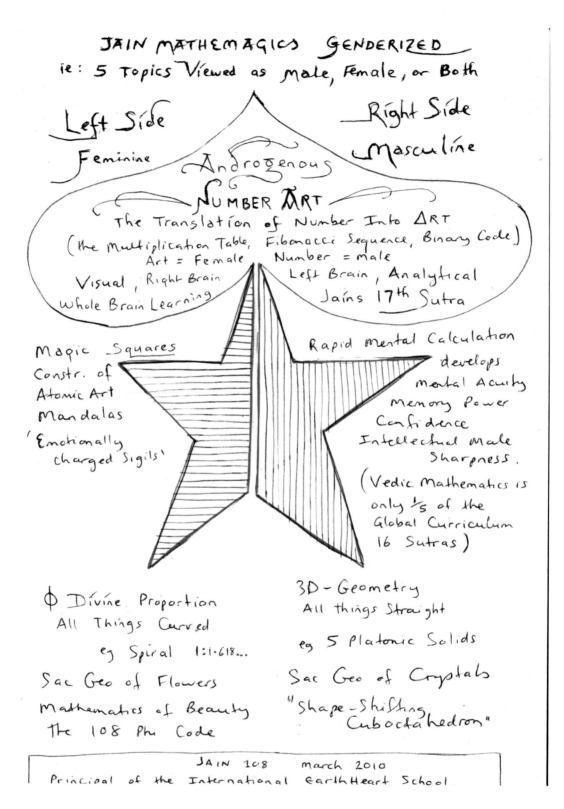

JAIN MATHEMAGICS GENDERIZED
ie: 5 Topics Viewed as Male, Female, or Both

Left Side
Feminine

Right Side
Masculine

Androgenous
NUMBER ART
The Translation of Number Into ART
(the Multiplication Table, Fibonacci Sequence, Binary Code)
Art = Female   Number = male
Visual, Right Brain   Left Brain, Analytical
Whole Brain Learning   Jains 17th Sutra

Magic Squares
Constr. of
Atomic Art
Mandalas
'Emotionally
charged Sigils'

Rapid Mental Calculation
develops
Mental Acuity
Memory Power
Confidence
Intellectual Male
Sharpness.
(Vedic Mathematics is
only ⅕ of the
Global Curriculum
16 Sutras)

Φ Divine Proportion
All Things Curved
eg Spiral 1:1·618...
Sac Geo of Flowers
Mathematics of Beauty
The 108 Phi Code

3D – Geometry
All things Straight
eg 5 Platonic Solids
Sac Geo of Crystals
"Shape-Shifting
Cuboctahedron"

JAIN 108   march 2010
Principal of the International EarthHeart School

This badge for Jain 108's International EarthHeart School illustrates in its 5 main arms, branches or rays **The QUINCUNX** Curriculum.

JAIN MATHEMAGICS CURRICULUM for the GLOBAL SCHOOL

formerly known as: The VEDIC MATHEMATICS CURRICULUM for the GLOBAL SCHOOL

$$13 \times 14 = 13 + 4 / 3 \times 4 = 17 / 12 = 18/2 = 182$$

$$3^3 + 4^3 + 5^3 = 6^3$$

$$98 \times 97 = 98 - 3 / 2 \times 3 = 95 / 06 = 9.506$$

Part 2 MULTIPLICATION   JAIN 2005
Developing the Inner Mental Screen

JAIN MATHEMAGICS CURRICULUM FOR THE GLOBAL SCHOOL part 2 MULTIPLICATION 2005

"THE FUTURE IN MATHEMATICS IS DOING IT IN YOUR HEAD" (by Jain: Mathematical Futurist).

• This is an Educational Workbook for Children and a Teacher's Resource Material for those interested in learning and teaching Rapid Mental Calculation. It increases the child's Memory Power and Confidence.

• This long-waited for book of Ancient Mathematical Short-Cuts excavates many hidden truths and vital properties in the playing field of Numbers.

• There is a rare and unique chapter called "HARMONIC STAIRWAY" from Jain's Dictionary of Numbers, for the first time in print. These choice Mathematical Plums are something that the student stumbles upon. They are working away solving calculations mentally, and when they check their answers in Chapter 2, they are instructed, in the section: DID YOU KNOW, to learn more about that particular number. But these are not any particular numbers, they are Anointed Numbers that our forebears held in high esteem.

• Beautifully and richly illustrated that the graphics alone educate. The student learns about other Families of Numbers, other numerical relatives and mathematical cousins that enrich their understanding of what mathematics is really about. It instills that sense of joy and wonderment that Pythagoras and Baudahayana knew.

• There is no error; it is an infallible bulletproof system, based on Unity Consciousness. The 16 "Threads" or Sutras that solve all known mathematical problems express The Law of Economy and The Path Of Least Resistance.

• As Jain Mathemagics becomes more globally acknowledged, it will help end the generational tyranny that has kept such knowledge in the dusty cupboard. This book is an organic pill that will prevent the slowly encroaching borgificiation of Mathematics.

• Since the turn of the bi-Millennium, there has been a global renaissance in the subject of "Sacred Geometry". You could summarize it as a fascination for the Language of Shape. This book invites seekers to truly incorporate SHAPE into their ability to perform mathematics, to apply this Language of Shape to the next octave of learning, not theoretically, but by practical use of using SHAPE to literally perform mental calculations in seconds.

• Often students come out of school unable to recover from deep mathematical wounds. They are mis-diagnosed as "un-intelligent" or "dumb", given drugs like Ritalin, told they are ADHD (which really means: Attuned Directly to Higher Dimensions). In fact, Dyslexic children are geniuses.

• **UNCOVER, RECOVER, DISCOVER**

(Front and Back Cover of my main book on Rapid Mental Calculation)

TRIBUTE TO LYNDON LAROUCHE

I would like to make a tribute, a thank you to Lyndon LaRouche. Whatever success I have had, is a demonstration of the true principle that should guide education and the development of human beings, a principle about which LaRouche has been speaking for years.

**"Obviously,
the difference between a beast and a man,
the characteristic difference,
is this quality of cognition, quality of reason.
The quality of making fundamental discoveries
which can be proven to be true about the universe.
So, our job is essentially to take a young child;
and, knowing in the child there is the spark of the ability
to make creative discoveries our job is to enable that child
to experience the great discoveries of principle of past
civilizations, and to embody those discoveries
in themselves."**

(Teach the "Eureka!" Principle Lyndon LaRouche May 24, 2001) Although we did not have a culture that provided the opportunity for us to do this as children, LaRouche created the conditions for us to do this as young adults. And so, he created the possibility for us to show that he is right. And indeed: LaRouche is right.

Works Cited:

Kepler, Johannes. *The Harmony of the World*. Trans. E.J. Aiton, et. al. American Philosophical Society, 1997

Kepler, Johannes. *Mysterium Cosmigraphicum*. Trans. A.M. Duncan. Abaris Books, Inc. New York, NY, 1981.

EXTRA PUZZLES AND LESSONS:

COMBINATORICS
&
The THEORY OF PROBABILITY

Some examples concerning the mathematics of options or possibilities are discussed under the following topics:

– rolling dice
– telephone numbers
– license plates
– lottery tickets
– line-ups in joint-photography
– triple-tiered ice-cream cones
– getting 4 aces in poker
– having 3 baby girls
– flipping coins to get 3 heads
– etc).

Utilizing the Multiplication Principle, for some choices "M" different ways and some choices that can be made in "N" different ways, then there are **M x N** various options or choices that can be made in succession.

| **EXAMPLE 1** |

If a woman has 5 blouses and 3 skirts, how many different ways can she dress with them?

EXAMPLE 2

From a menu that has 4 appetizers, 7 main meals and 3 desserts, how many different dinners can made?

EXAMPLE 3

We know that when rolling dice there are 6 x 6 = 36 possible outcomes for any two number combinations from 1 to 6. It is to be noted that these 36 possibilities do include repetitions of the same number like 3 and 3.

But, the question is, how many possible outcomes are there when no repetitions are allowed, ie: the second dice has a different value than the first dice?

EXAMPLE 4

a) − How many possible outcomes are there when rolling 3 die, allowing for any combination including repetitions?

b) − How many possible outcomes are there when rolling 3 die, without including repetitions?

EXAMPLE 5

Can you estimate how many differing telephone numbers can be reached , assuming:

  – there are 7 digits in each number,

  – there are no area codes,

  – that the first digit can not be a "0" or a "1" but the remaining digits can be occupied with a choice from 10 digits, ie: zero and the digits from 1 to 9. This means that there is only a choice of 8 digits in this first position! (being a 2-3-4-5-6-7-8 or 9).

EXAMPLE 6

How many various number plates can be made if the conditions were based on 2 Letters of the English Alphabet followed by 4 numbers?

EXAMPLE 7

a) – There are 8 people being jointly photographed at the end of a Jain and Lily seminar.
a) – How may different line-ups of people are possible?

b) – Out of this number of different line-ups, how many different ways can Jain and Lily, who want to be photographed standing next to one another, be achieved?
Here is a <u>Clue</u>: visualize that Jain and Lily are actually one entity, so think of "7 Factorial".

c) – What is the probability (expressed in terms of a simple fraction like 1/3) that Jain and Lily can be photographed next to one another?

EXAMPLE 8

a) – The Lily-Jain Organic IceCream Shop in Byron Bay advertises 24 different flavours of ice-cream.
a) – What is the number of possible triple-scoop cones, not including any repetitions of flavours?

b) – If we are not interested in how the flavours are arranged, how many 3-flavoured cones are possible?
<u>Clue</u>: establish first how many different ways there are to arrange the 3 flavours!

EXAMPLE 9

a) – How many different ways are there in expressing the 6 required numbers of a lottery ticket when there is a pool or choice of 40 different numbers (1 to 40), assuming we are interested in the specific order of these 6 numbered sequences. ie: What are your chances to be a winner in this lottery, by choosing the 6 numbers out of the possible 40?

b) – If we are not interested in the order of the 6 numbered sequences, as is the case with general lotteries, meaning any of the 6 numbers contributes to your chances of winning, then what are number of possible arrangements?
<u>Clue</u>: Firstly work out the different number of possibilities in any 6 numbered sequence!

EXAMPLE 10

a) – How many different poker hands of 5 cards are possible to be dealt, assuming the order of the cards is important?

b) – How many possible ways are there of being dealt 4 Aces?

HEADS AND TAILS:

Such Numbers calculated in the previous 10 examples come under the banner of Combinatorial Coefficients which observes the number of ways of choosing M elements out of N elements, with the condition that we are not interested in the order of these R elements, as shown in Example 8:

$(24 \times 23 \times 22) / (3 \times 2 \times 1)$.

An analogue (something having analogy to something else) to this MxN Multiplication Principle is used in the Theory of Probability. Have a look at this simple example:

EXAMPLE 11

What is the probability of getting two Heads in the flip of a Coin that has Heads (H) and Tails(T)?

EXAMPLE 12

Using one coin, what are the chances of getting HH in 10 successive or consecutive flips of the coin?

EXAMPLE 13

If we simplify the number of days in a year to 366, and we are interested in the search for two people who have the same birthday, then how many people have to be in the same hall to mathematically predict or be absolutely certain that two people in this hall have the same birthday?

EXAMPLE 14

Tests have been done to observe say 1000 married couples, tracking or observing them for ten years, with the specific intention for them to acquire 3 children. 800 couples do manage to have 3 children.

a) – What is the chance or probability of having 3 Girls?
ie: of the 800 couples, how many couples have 3 Girls?

b What is the chance of having 2 Girls (G) and a Boy (B)?

ANSWER 1

There are 5 x 3 different ways = 15 options.

ANSWER 2

There are 4 x 7 x 3 = 84 different dinners.

ANSWER 3

There are 6 x 5 = 30 possible outcomes. This is because the first dice can combine with any of the remaining 5 dice.

ANSWER 4

a) – There are 6 x 6 x 6 = 216 possibilities.
b) – There are 6 x 5 x 4 = 120 possibilities.

ANSWER 5

There are 8×10^6 or 8 million possible phone numbers that can be rung.

There will be 26^2 x 10^4 = 6,760,000 possible number plates.

ANSWER 7

a) – There are 8 x 7 x 6 x 5 x 4 x 3 x 2 x 1 = 40,320 different ways.
nb: another way of expressing he multiplication of all the numbers from 1 to 8, as shown above, is called "8 Factorial".

b) – There are 7 x 6 x 5 x 4 x 3 x 2 x 1 = 5,040 possible ways, but we need to multiply this result by 2, because there are two options in how Jain and Lily stand together, as in Jain is first, then Lily is next, and vice-versa, so this gives a result of 5,040 x 2 = 10,080 possible ways that Jain and Lily can be standing next to one another.

c) – Thus if Jain and Lily are randomly lined up, the probability that both of them will be standing next to one another is 10,080 divided by 40,320 = ¼.

ANSWER 8

a) – There are 24 x 23 x 22 = 12,144 various 3 tiered flavours. This allows for the fact that 24 flavours can be on top, then any of the remaining 23 can be placed in the middle, and any of the remaining 22 can be placed on the bottom. nb: this is a different to the idea of calculating the possibility of 3-flavoured cones if we are not interested in how the flavours are arranged!

b) – Firstly, there are 3 x 2 x 1 = 6 different ways to arrange the 3 flavours. eg: Mango Boysenberry Coconut is represented as MBC and this arrangement can be coded as MCB, BMC, BCM, CMB and CBM giving a total of 6 permutations.
Therefore we need to divide our number of 12,144 3-tiered flavours by 6 giving a result of 2,026 ice-cream cones.
Mathematically, you would write the above data as:
(24 x 23 x 22) / (3 x 2 x 1) = 2,026.

ANSWER 9

a) – There are (40 x 39 x 38 x 37 x 36 x 35) = 2,763,633,600 or approximately 2.8 billion possible ways or choices, if we are interested in the specific order of these 6 numbered sequences.

b) – Firstly, the different number of possibilities in any 6 numbered sequence is arrived at by working out "6 Factorial" which is:
6 x 5 x 4 x 3 x 2 x 1 = 720 possibilities.
We then have to divide 2,763,633,600 by 720 giving 3,838,380 possible 6 numbered sequences.

ANSWER 10

a) – There are (52 x 51 x 50 x 49 x 48) possible poker hands that can be dealt, if the order is relevant, but since it is not relevant in this case, we need to divide the above products by "5 Factorial" giving a result of:
(52 x 51 x 50 x 49 x 48) / (5 x 4 x 3 x 2 x 1) = 2,598,960 possible poker hands.

b) – There are 48 / 2,598,960 possible ways of being dealt 4 Aces, which reduces to about 1 in 50,000. This is because there are 48 different ways of being dealt a poker hand with 4 aces understanding that these 48 cards could be the 5th card in such a poker hand.

ANSWER 11

The 4 possibilities are HH, HT, TH, TT therefore the probability is 1 in 4 or ½ x ½ = ¼.

ANSWER 12

There is a $(½)^5$ = 1/32 result.

There must be 366 + 1 = 367 people in the hall to guarantee that at least 2 people have the same birthday.

| ANSWER 14 |

a) – It is a ½ x ½ x ½ = 1/8 chance of having 3 girls. This means 100 of the 800 do have 3 girls.

b) – 300 of the 800 have GGB or 3/8. This is because if 100 couples have GGG then it is understood by the Law of Balance that another 100 couples must have BBB.
This leaves an amount of 800 – 100 – 100 = 600 couples.
Let us look at the 3 different sequences possible for Two Girls and One Boy, they are:
BGG, GBG and GGB.
Since this sequence of 3 possibilities is the same for the sequence of Two Boys and One Girl, then the answer must be ½ of 600 = 300.

"**I see a Future where all Children of the World are doing their Mathematics swiftly in their Head**"

by Jain 108,
Mathematical Futurist and
EcoPreneur,
(in contrast to "EntrePreneur"
which, as you sadly know,
literally means: "Enter and Take"!)

Mural by Jain 1999 "Dolphusion".
Thanks to Aaron Lutze of the Gold Coast, for his computer skills designing this dvd cover.

APPENDIX

1895 the 8th GRADE FINAL EXAM

What it took to get an 8th grade education in 1895...

Remember when grandparents and great-grandparents stated that they only had an 8th grade education? Well, check this out. Could any of us have passed the 8th grade in 1895?

This is the eighth-grade final exam from 1895 in Salina, Kansas, USA. It was taken from the original document on file at the Smokey Valley Genealogical Society and Library in Salina, and reprinted by the Salina Journal.

8th GRADE FINAL EXAM: Salina , KS - 1895

Grammar (Time, one hour)
1. Give nine rules for the use of capital letters.
2. Name the parts of speech and define those that have no modifications.
3. Define verse, stanza and paragraph
4. What are the principal parts of a verb? Give principal parts of 'lie, ''play,' and 'run.'
5. Define case; illustrate each case.
6. What is punctuation? Give rules for principal marks of punctuation.
7. Write a composition of about 150 words and show therein that you understand the practical use of the rules of grammar.

Arithmetic (Time, 1 hour 15 minutes)
1. Name and define the Fundamental Rules of Arithmetic.
2. A wagon box is 2 ft. Deep, 10 feet long, and 3 ft. Wide. How many bushels of wheat will it hold?
3. If a load of wheat weighs 3,942 lbs., what is it worth at 50cts/bushel, deducting 1,050 lbs. For tare?
4. District No 33 has a valuation of $35,000.. What is the necessary levy to carry on a school seven months at $50 per month, and have $104 for incidentals?
5. Find the cost of 6,720 lbs. of coal at $6.00 per ton.
6. Find the interest of $512.60 for 8 months and 18 days at 7 percent.
7. What is the cost of 40 boards 12 inches wide and 16 ft. long at $20 per yard?
8. Find bank discount on $300 for 90 days (no grace) at 10 percent.
9. What is the cost of a square farm at $15 per acre, the distance of which is 640 rods?
10. Write a Bank Check, a Promissory Note, and a Receipt.

U.S. History (Time, 45 minutes)

1. Give the epochs into which U.S. History is divided.
2. Give an account of the discovery of America by Columbus.
3. Relate the causes and results of the Revolutionary War.
4. Show the territorial growth of the United States.
5. Tell what you can of the history of Kansas.
6. Describe three of the most prominent battles of the Rebellion.
7. Who were the following: Morse, Whitney, Fulton, Bell, Lincoln, Penn and Howe?
8. Name events connected with the following dates: 1607, 1620, 1800, 1849, 1865.

Orthography (Time, one hour)

1. What is meant by the following: alphabet, phonetic, orthography, etymology, syllabication
2. What are elementary sounds? How classified?
3. What are the following, and give examples of each: trigraph, subvocals, diphthong, cognate letters, linguals
4. Give four substitutes for caret 'u'.
5. Give two rules for spelling words with final 'e.' Name two exceptions under each rule.
6. Give two uses of silent letters in spelling. Illustrate each.
7. Define the following prefixes and use in connection with a word: bi, dis-mis, pre, semi, post, non, inter, mono, sup.
8. Mark diacritically and divide into syllables the following, and name the sign that indicates the sound: card, ball, mercy, sir, odd, cell, rise, blood, fare, last.
9. Use the following correctly in sentences: cite, site, sight, fane, fain, feign, vane, vain, vein, raze, raise, rays.
10. Write 10 words frequently mispronounced and indicate pronunciation by use of diacritical marks and by syllabication.

Geography (Time, one hour)

1. What is climate? Upon what does climate depend?

2. How do you account for the extremes of climate in Kansas?

3. Of what use are rivers? Of what use is the ocean?

4. Describe the mountains of North America.

5. Name and describe the following: Monrovia, Odessa, Denver, Manitoba, Hecla, Yukon, St. Helena, Juan Fernandez, Aspinwall and Orinoco.

6. Name and locate the principal trade centers of the U.S.

7. Name all the republics of Europe and give the capital of each.

8. Why is the Atlantic Coast colder than the Pacific in the same latitude?

9. Describe the process by which the water of the ocean returns to the sources of rivers.

10. Describe the movements of the earth. Give the inclination of the earth.

✶✶✶✶✶✶✶✶✶✶✶✶✶✶

Notice that the exam took FIVE HOURS to complete.
I don't even know what "Orthography" means!?"

Gives the saying "he/she only had an 8th grade education" a whole new meaning, doesn't it?!

No wonder they dropped out after 8th grade. They already knew more than they needed to know!

✶✶✶✶✶✶✶✶✶✶✶✶✶✶

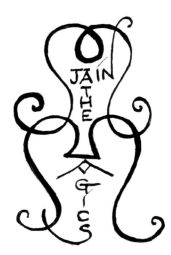

nb: all numbers in bold refer to that reference being an image.

O
Odd Numbers: 11, **20**, 51, 55-**58**-59,
One-Line Answer: 101-103,

P
Pairs of 10: 80,
Papaya: **47**,
Party Trick: 83,
Pattern Recognition: 55, 73,
Pascal's Triangle: 11-20, 140,
Pascal, Blaise: **11**,
Pentagon: 128 (3-4-5 Triangle),
Percentages: 21-29, 140,
Periodicity: 101-103,
Phi Ratio 1:1.618033...: 10, 114, 123, 130, 133,
Place Value System: 33, 75,
Platinum crystal: 10,
Platonic Solids: 10,
Powers of 11: 11, 13,
Prime Number Cross: **2**, 101-103,
Probability Theory: 152-162,
Puzzle: **131-132**,
Pythagoras: 58,
Pythagoras Theorem: 114-134,
   p**116** (The 108 Triangle),
   p**117-119**, 143, (Pythagorean Triples)
   p**120-121** (The Rope-Stretchers Knot of 12),
   p**120** (Gnomon)
   p**121** (The 3-4-5 Triangle In the Geometry of Flowers)
   p**123** (Plato's "Most Beautiful Triangle")
   p**122** (3-4-5 Proof by Mere Observation from Greece)
   p**122** (3-4-5 Algebraic Proof by Bhaskara, India
   p124-128 (Some Ancient Historical References on the
         3-4-5 Triangle)
   p**128** (Pythagorean derivation of the Musical
         Diatessaron or 3:4 Ratio)
   p**129** (Derivation of the Symbols of the Master Mason)
   p**130-134**, 143, (Using Pythagoras' Theorem to
         determine $\sqrt{2}$, $\sqrt{3}$ and $\sqrt{5}$ in the Vesica Piscis)
   p**131-132** (Puzzle using Pythagoras' Theorem to
         determine the Shortest Path on The Cube)

RAPID MENTAL CALCULATION & SACRED GEOMETRY

Private Tutoring

$108 = 1^1 . 2^2 . 3^3$

$$\text{Phi } (\phi) = \frac{1 + \sqrt{5}}{2}$$
$$= 1.618033988...$$

for
Teenagers

JainMathemagics.com

Poster advertising Private Mathematics Tutoring offered by Jain
in the Mullumbimby, Byron Bay Shire, far north NSW, 2011.
(Shown above: Mingkah Jain Sun on the left and Aysha Jain Sun).

The End

Made in United States
Orlando, FL
14 July 2024

48937320R00100